Rudolph /
Understanding Plastics Recycling

Natalie Rudolph
Raphael Kiesel
Chuanchom Aumnate

Understanding Plastics Recycling

Economic, Ecological, and Technical Aspects of
Plastic Waste Handling

3rd Edition

HANSER

Print-ISBN: 978-1-56990-865-5
E-Book-ISBN: 978-1-56990-945-4
E-Pub-ISBN: 978-1-56990-957-7

Bibliographic information of the German National Library:
The German National Library lists this publication in the German National Bibliography; detailed bibliographic data are available on the Internet at http://dnb.d-nb.de.

© 2025 Carl Hanser Verlag GmbH & Co. KG, Munich
Vilshofener Straße 10 | 81679 Munich | info@hanser.de
www.hanserpublications.com
www.hanser-fachbuch.de
Editor: Dr. Mark Smith
Production Management: Cornelia Speckmaier
Cover concept: Marc Müller-Bremer, www.rebranding.de, Munich
Cover design: Max Kostopoulos
Cover picture: © istockphoto.com/Kriengsak Tarasri; Icons: flaticon.com
Typesetting: Eberl & Koesel Studio, Kempten

Contents

Foreword

The plastics industry is undergoing a profound transformation. In the face of increasing regulatory pressure, global sustainability goals, and shifting customer expectations, producers are rethinking how plastics are designed, produced, and recycled. By 2050, European plastics producers aim to replace 65% of our fossil feedstock with circular or renewable feedstock and decarbonize their production processes.

With the "Plastics Transition" initiative, Plastics Europe has united its members – leading plastics producers and chemical companies – around a shared vision for a climate-neutral, circular plastics system. This roadmap identifies key enablers such as circular product design, scalable chemical recycling, the shift to low carbon energy sources, and greater transparency across the entire value chain. While these goals are ambitious, they reflect the industry's shared commitment to systemic change.

But policy ambitions and industrial roadmaps must be supported by technical knowledge and practical tools. This is where *Understanding Plastics Recycling* by Natalie Rudolph, Raphael Kiesel, and Chuanchom Aumnate offers important guidance. The book provides a well-structured introduction to the key aspects of plastics recycling. It combines technical foundations with economic and regulatory perspectives, making it relevant not only for students, but also for professionals in industry, academia, and policy.

As demand for recycled plastics continues to grow, so do the requirements for quality, traceability, and performance. Particularly in sensitive application areas such as packaging, technical components, or agricultural films, the use of recyclates involves complex technical, economic, and regulatory considerations. This book helps readers understand the nuances and apply them in practice.

It covers various mechanical and chemical recycling processes, with a particular focus on automated sorting and processing of plastic waste. Section 2.3 outlines a range of sorting technologies, from infrared-based systems to AI-assisted and tracer-based approaches, and illustrates how sorting and processing have evolved in recent years.

Chapter 3 is dedicated to quality assurance. It provides context for testing methods such as thermal analysis, rheology, or spectroscopic techniques, and explains their relevance for the use of recycled plastics in higher-value applications.

The economic and environmental framework conditions for recycling are also addressed. Chapter 4 discusses energy use, feedstock availability, and carbon footprints for both virgin and recycled materials. It explores the boundaries of economic and ecological viability and offers a realistic view of current trade-offs.

An important section of the book is devoted to legislation. Chapter 5 examines the role of political and regulatory measures in shaping recycling markets, including national and European-level targets, as well as questions of international harmonization.

Understanding Plastics Recycling is a timely and practical resource for anyone looking to deepen their understanding of circularity in the plastics sector. It bridges the gap between vision and implementation and supports the fact-based dialogue needed across the value chain – from producers and converters to policymakers and recyclers.

I wish all readers an engaging and insightful read.

Dr. Christine Bunte
Managing Director
Plastics Europe Deutschland

Plastics Europe is the pan-European association of plastics manufacturers with offices across Europe. For over 100 years, science and innovation has been the DNA that cuts across our industry. With close to 100 members producing over 90% of all polymers across Europe, we are the catalyst for the industry with a responsibility to openly engage with stakeholders and deliver solutions which are safe, circular, and sustainable. We are committed to implementing long-lasting positive change.

The Authors

The first edition of Understanding Plastics Recycling resulted from our time together at the University of Wisconsin–Madison in 2015–2016. There, we worked side by side in research and public outreach on the topic of plastics recycling. That period was marked by a shared drive to advance technical understanding and improve the public conversation about sustainability and circularity in the United States. This collaboration laid the foundation for the first edition of this book, and our friendship has kept us going through frequent updates – as the recycling industry continues to evolve rapidly alongside advances in materials science and processing technologies.

Dr. Natalie Rudolph

Natalie Rudolph holds a Ph.D. in Polymer Engineering and has worked in both academia and industry with a focus on polymer materials, processing, and analysis. After completing her doctorate in Germany, she moved to the United States for a postdoctoral position at the University of Wisconsin–Madison.

She later served as Department Head for Materials and Testing at the Fraunhofer Institute for Chemical Technology in Germany, where her work centered on lightweight materials and their testing, contributing to resource efficiency and sustainability in mobility and energy applications.

In 2014, she returned to the University of Wisconsin–Madison as an Assistant Professor of Mechanical Engineering and Associate Director of the Polymer Engineering Center. During this time, she conducted research on additive manufacturing, composite materials, and plastics recycling. It was here that the collaboration with her co-authors began.

Following her academic role, she transitioned into industry leadership, including her position as Vice President of R&D at AREVO Inc., where she helped scale up hardware and software for composite 3D printing aimed at enabling local, automated manufacturing – supporting more sustainable production models.

Since 2020, she has been with NETZSCH Analyzing & Testing in Germany, currently serving as Division Manager Polymer. In this role, she is responsible for global business development, application strategy, and technical communication related to thermal analysis and rheology in the polymer industry. A key focus of her current work is enabling practical tools for circular economy applications through advanced material analysis.

Dr. Raphael Kiesel

Holding both a Ph.D. in Mechanical Engineering and MBA, Raphael Kiesel combines knowledge in the fields of production technology, digitalization, and sustainability with an entrepreneurial mindset. His early research at the Polymer Engineering Center in Wisconsin centered on plastics recycling, laying the foundation for his ongoing commitment to circular economy solutions.

He went on to lead research groups at the Fraunhofer IPT and WZL Institutes, where he focused on data-driven quality management and the application of AI and 5G technologies for sustainable manufacturing. As Head of Department, he managed interdisciplinary teams and coordinated national and international industry collaborations.

Beginning of 2023, Raphael Kiesel joined ARRI Group, first as Senior Vice President Quality Management and then as Senior Vice President Business Unit Lighting. In these roles, he was overseeing strategy, product development, operations, and post-merger integration.

In September 2025, he assumed the role Head of Corporate Strategy at Rohde and Schwarz, where he combines his technology know-how with his entrepreneurial thinking.

Alongside his industry roles, he lectures on smart manufacturing and is a speaker on the topics of circularity and digitalization. A member of the Think Tank 30 of the Club of Rome, Raphael works at the intersection of technology, business, and sustainability – driven by the vision of enabling circularity through innovation.

Dr. Chuanchom Aumnate

Dr. Chuanchom Aumnate holds a Ph.D. in Mechanical Engineering with a specialization in polymer processing and rheology. After completing her doctoral studies at the University of Wisconsin–Madison, she returned to Thailand to begin her research career at the Metallurgy and Materials Science Research Institute, Chulalongkorn University.

By 2021, she had advanced to an expert-level research position at the same institute, where her work centered on sustainable polymers, recycled plastics, advanced composites, and additive manufacturing.

Dr. Aumnate is currently a faculty member at The Petroleum and Petrochemical College (PPC), Chulalongkorn University. She actively contributes to both teaching and research, with a particular focus on polymer recycling, structure–property relationships in recycled materials, and the integration of renewable fillers and nanomaterials to develop high-performance, value-added products. Her research applies polymer rheology, thermal analysis, and 3D printing technologies to create innovative materials for circular economy applications.

She has extensive experience in collaborative research with both domestic and international partners. Her work has significantly advanced the understanding of how to enhance the mechanical integrity, processability, and environmental performance of bio-derived materials and recycled plastics through material design and process innovation.

Acronyms and Other Abbreviations

Abbreviation	Description
ABS	acrylonitrile butadiene styrene
ARR	average rate of return
ASTM	American Society for Testing and Materials
CCM	cost comparison method
CLF	closed loop fund
DSC	differential scanning calorimetry
EPA	U.S. Environmental Protection Agency
EPS	expanded polystyrene
GHG	greenhouse gas
HDPE	high-density polyethylene
HIPS	high-impact polystyrene
LDPE	low-density polyethylene
LFG	landfill gas
LLDPE	linear low-density polyethylene
MFI	melt flow index
MFR	melt flow rate
MRF	materials recovery facility
MSW	municipal solid waste
OCC	old corrugated cardboard

Abbreviation	Description
PA	polyamide
PBT	polybutylene terephthalate
PC	polycarbonate
PCM	profit comparison method
PE	polyethylene
PEEK	polyether ether ketone (or polyarylether etherketone)
PET	polyethylene terephthalate
PLA	polylactide
PMMA	polymethyl methacrylate
POM	polyoxymethylene (polyacetals)
PP	polypropylene
PPE	polyphenylene ether
PPP	purchasing power parity
PRF	plastics recycling facility
PS	polystyrene
PTFE	polytetrafluoroethylene
PU	polyurethane
PVC	polyvinyl chloride
QA/QC	quality assurance/quality control
RCRA	Resource Conservation and Recovery Act
rLDPE	recycled low-density polyethylene
RoM	rule of mixtures
rPP	recycled polypropylene
SAN	styrene acrylonitrile
SNCR	selective noncatalytic reduction
SPP	static payback period
UV	ultraviolet
WARM	EPA Waste Reduction Model
WTE	waste-to-energy
XPS	extruded polystyrene

1

Circular Economy of Plastics

"For the last 150 years, plastic materials have been key enablers for innovation and have contributed to the development and progress of society." [1]. This quote from PlasticsEurope portrays the importance of plastics in today's industrialized society. Imagining today's world without plastics is virtually impossible. Compared to every other engineered material in the world, plastics have the highest growth rate, which is related to their unique properties. Plastics have become the most important raw material for a variety of products and applications. This includes healthcare (medical devices, PPE), packaging (food preservation), construction, electronics, and transportation. Their versatility, durability, high strength-to-weight ratio, and cost-efficient manufacturing methods result in many benefits [2, 3]. For example, plastics are critical in reducing food waste, with plastic packaging extending shelf life by 2–3 times [4, 5]. Furthermore, plastic-based components have reduced the weight of modern vehicles, improving fuel efficiency by 25–30% [6].

These benefits led to a twentyfold exponential growth in their production in the past half-century. Between 2010 and 2023, worldwide plastics production increased by over 50% from 270 Mt to 414 Mt [7]. By 2030 their production is expected to increase to 600 Mt [8].

Increasing production consequently results in higher amounts of plastic waste. In 2024, 360 Mt of plastic waste were generated worldwide, with packaging waste accounting for the highest amount (156 Mt) [9, 10]. Figure 1.1 shows the plastic waste distribution of the different geographic regions.

The growing volume of plastic waste, combined with increasing environmental awareness, highlights the urgency of addressing plastic pollution: Every minute, the equivalent of a garbage truck's worth of plastic enters our oceans; an estimated 8–12 million tons of plastic enter the oceans annually [13]. Furthermore, carbon emissions from plastic production are projected to contribute to 19% of global carbon budget by 2040 [14]. Plastic waste generation is projected to triple by 2060 if current trends continue, which increases the described environmental problems [15].

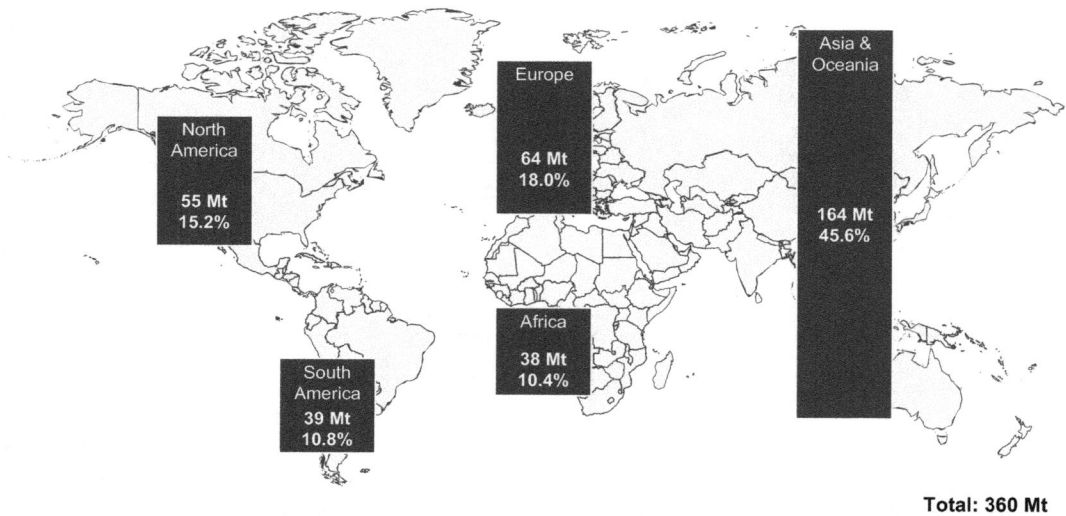

Figure 1.1 Postconsumer plastic waste worldwide in 2024 (Remark: Europe covers both EU and non-EU countries) [10, 11, 12]

However, despite the growing amount of plastic waste, the share of recycled plastics does not significantly increase. In 2022, only 9% of plastic waste was recycled globally. The remaining waste was predominantly sent to landfills (49%), incinerated (19%), or mismanaged (22%) [15].

Thereby, circular strategies – including plastics recycling – are key for the environment. Circular strategies are projected to reduce plastic waste by up to 80%. By keeping plastics in a closed loop, circular strategies prevent plastics from entering landfills, waterways, and ecosystems. This minimizes the harmful effects of microplastics on marine life and biodiversity. By reusing and recycling plastics, circular strategies reduce reliance on finite resources like fossil fuels. This enhances resource efficiency and reduces greenhouse gas emissions associated with virgin plastic production. Circular approaches can significantly lower greenhouse gas emissions – by up to 40% according to some estimates – helping industries align with global climate goals. Besides the environmental benefits, transitioning to a circular economy reduces material costs, enhances supply chain resilience, and creates new business opportunities. By 2040, these new business opportunities could save $200 billion annually and create 700,000 additional jobs globally [16, 17, 18, 19].

Therefore, Section 1.1 describes circular strategies for plastic waste, including their impact on circularity and an example. Section 1.2 focuses on recycling, as this has the highest impact in the next decades due to all plastics already produced.

1.1 9R Framework to Prevent and Manage Plastic Waste

	STRATEGY	DESCRIPTION	IMPACT	EXAMPLE
Smarter product use and manufacture	R0 | Reuse	Avoid plastic use altogether by declining unnecessary products or packaging.	Reduces the demand for new plastic production.	Avoid single-use plastic bags by using reusable alternatives.
	R1 | Rethink	Reimagine product design to minimize plastic content or make it reusable or durable.	Enables more sustainable product use patterns.	Designing products with modular, repairable components to extend their lifespan.
	R2 | Reduce	Minimize plastic consumption using less material or replacing it with alternatives.	Reduces resource extraction and waste generation.	Lightweight packaging or using biodegradable materials where appropriate.
Extend lifespan of product and its parts	R3 | Reuse	Extend the life of plastic products by using them multiple times.	Decreases the need for new plastic production and delays waste generation.	Implementing refillable containers for food or personal care products.
	R4 | Repair	Fix damaged plastic items to extend their use.	Reduces waste by keeping products in use longer.	Repairing durable goods like plastic furniture or automotive components.
	R5 | Refurbish	Restore used items to a functional condition, often involving multiple components.	Reduces the need for both new materials and disposal of old items.	Refurbishing plastic office equipment or electronic devices with plastic parts.
	R6 | Remanufacture	Rebuild products with used parts or materials.	Recirculates materials in the economy and conserves resources.	Remanufacturing car parts that include plastic components.
	R7 | Repurpose	Use plastic waste in new applications without reprocessing into raw materials.	Gives materials a second life, delaying waste disposal.	Using old plastic barrels as planters or storage containers.
Useful application of materials	R8 | Recycle	Process plastic waste into raw materials for new products.	Reduces demand for virgin plastic; lower energy consumption than new production.	Turning PET bottles into new packaging or polyester fibers.
	R9 | Recover	Energy recovery through incineration if plastics can no longer be recycled or reused.	Reduces landfill waste but releases carbon emissions and other pollutants.	Incinerating plastic waste to generate electricity.

Increasing Circularity →

Figure 1.2 9R Framework prioritizing the circular economy strategies according to their levels of circularity [20]

Various approaches, known as R-strategies, have been developed to achieve less resource and material consumption in product chains and make the economy more circular. The 9R Framework presents 10 circular strategies ordered from high circularity (low R-number) to low circularity (high R-number). A higher level of circularity of materials in product chains means that, in principle, smaller amounts of natural resources are needed to produce new (primary or virgin) materials. Therefore, whenever possible, strategies with lower numbers should be prioritized to minimize waste and maximize resource efficiency [20, 21].

Figure 1.2 presents the 9R Framework as applied to plastics. It outlines each R-strategy, describing its general concept, its specific impact on plastics, and providing a relevant example for each strategy.

1.2 Recycling – The Key for Circular Economy of Plastics

The application of reduction and reuse strategies for plastics is the most favorable in the context of circular economy. However, for plastics specifically, there are certain limits when it comes to reduction and reuse. As described in the introduction of this chapter, plastics have become the most important raw material for a variety of products and applications and are partially irreplaceable. Not all plastic can be eliminated without compromising safety, hygiene, or efficiency (e. g., medical syringes, food preservation). Furthermore, many plastic products are designed for single use, making reuse impractical or unsafe. Packaging in the food industry is often non-reusable due to contamination risks.

If plastic waste cannot be prevented, three ways of handling plastic waste exist, which are landfilling, incineration with energy recovery (waste-to-energy – WTE), and recycling. Compared to landfilling and WTE, recycling conserves resources, reduces energy consumption, and minimizes carbon emissions compared to producing new plastics [22].

The general cyclic recycling process involves three main steps:

1. Collecting the recyclables

2. Processing the recyclables (mechanical or chemical)

3. Turning the recyclables into new products

After collecting the recyclables via curbside collection, drop-off programs, buyback operations, and container-deposit systems, they are transported to material recovery facilities (MRFs), mixed-waste processing facilities, or mixed-waste composting facilities [23, 24]. High-tech MRFs are characterized by the automated separation of unsorted recyclables using eddy currents, magnetic pulleys, optical sensors, and air classifiers, reducing manual sorting, see Section 2.3.2. Automatic sorting supports and simplifies recycling and enhances its economic profitability [23].

Despite this necessity for recycling, only 9% of plastic waste is recycled globally. There are three major reasons for this: *contamination, downcycling,* and *sorting.* First, the contamination of mixed plastics and impurities hinder recycling efficiency. Second, recycling lowers the quality of plastics and their properties, limiting their applications. Third, sorting of certain plastic types can be technically challenging or is economically not feasible yet.

To close the loop and achieve the advantages of circular economy and plastics recycling, new recycling technologies, including chemical recycling, offer potential solutions to both increase the recycling rates and the economics behind the process, but require further development. Furthermore, effective regulations (e. g., extended producer responsibility, waste sorting mandates) can boost recycling rates. Countries with advanced recycling policies, like Germany, have achieved plastic recycling rates of 68% for packaging waste in 2023 [25].

Therefore, this book summarizes in Chapter 2 the most relevant plastic recycling techniques, as well as plastic characterization methods (Chapter 3), analyzes both economic and environmental aspects of different recycling methodologies (Chapter 4), gives insights into global policies and processes (Chapters 5 and 6), and gives an outlook regarding the future of plastics recycling (Chapter 7).

References

[1] PlasticsEurope. *Plastics – The Facts 2015: An analysis of European plastics production, demand and waste data*. PlasticsEurope. 2015. pp. 1–30.

[2] Niessner, N. *Recycling of Plastics*: Hanser: Munich. 2022.

[3] Baur, E., Drummer, D., Osswald, T. A., and Rudolph, N. *Saechtling Kunststoff-Handbuch: Eigenschaften, Verarbeitung, Konstruktion*: Hanser: Munich. 2022.

[4] Matthews, C., Moran, F., and Jaiswal, A. K. A review on European Union's strategy for plastics in a circular economy and its impact on food safety. *Journal of Cleaner Production*. vol. 283. 2021. 125263.

[5] Ncube, L. K., Ude, A. U., Ogunmuyiwa, E. N., Zulkifli, R., and Beas, I. N. An overview of plastic waste generation and management in food packaging industries. *Recycling*. vol. 6. 2021. p. 12.

[6] Bailo, C., Modi, S., Schultz, M., Fiorelli, T., Smith, B., and Snell, N. *Vehicle Mass Reduction Roadmap Study 2025–2035*. Center for Automotive Research: Detroit, MI. 2020.

[7] PlasticsEurope. *Plastics – the fast Facts 2024*. PlasticsEurope: Brussels. 2024.

[8] Statista. *Production forecast of thermoplastics worldwide from 2025 to 2050*. Access Date: 16.02.2025. Available: *https://www.statista.com/statistics/664906/plastics-production-volume-forecast-worldwide/#: ~:text=Plastic%20production%20forecast%20worldwide%202025%2D2050&text=Annual%20production %20volumes%20are%20expected,30%20percent%20compared%20with%202025*.

[9] Statista. *The World Is Flooded With Plastic Waste*. Access Date. Available: *https://www.statista.com/ chart/32385/global-plastic-waste-production-by-application/*.

[10] Statista. Projected plastic waste generation worldwide from 2020 to 2040. Access Date: 16.03.2025. Available: *https://www.statista.com/statistics/1559351/projected-plastic-waste-generation/*.

[11] Conversio Market & Strategy GmbH. *Global Plastic Flow 2018*. Mainaschaff, Germany. 2020.

[12] Kiesel, R. Comparison of Global Recycling Processes. *Recycling of Plastics*. Niessner, N., Ed. Hanser: Munich. 2022.

[13] Ellen Macarthur Foundation. *Plastics and the circular economy – deep dive*. Access Date: 16.02.2025. Available: *https://www.ellenmacarthurfoundation.org/plastics-and-the-circular-economy-deep-dive*.

[14] United Nations Climate Change. *A New Plastics Economy is Needed to Protect the Climate*. Access Date: 16.02.2025. Available: *https://unfccc.int/news/a-new-plastics-economy-is-needed-to-protect-the-climate#:~:text=Under%20a%20business%2Das%2Dusual,degrees%20Celsius%20out%20of%20reach*.

[15] OECD. *Global Plastics Outlook: Policy Scenarios to 2060*. Paris, France. 2022.

[16] AION. *The Circular Economy: A Sustainable Solution to the Plastic Crisis*. Access Date: 15.02.2025. Available: *https://www.aion.eco/resources/the-circular-economy-a-sustainable-solution-to-the-plastic-crisis*.

[17] Contec. *Plastics in a circular economy*. Access Date: 15.02.2025. Available: *https://contec.tech/euro pean-strategy-plastics-circular-economy/*.

[18] Kaizen Institute. *The Role of Circular Economy in the Chemical and Plastic Industry*. Access Date: 16.02.2025. Available: *https://kaizen.com/insights/circular-economy-chemical-plastic/*.

[19] Ellen Macarthur Foundation. *Designing out plastic pollution*. Access Date. Available: *https://www. ellenmacarthurfoundation.org/topics/plastics/overview*.

[20] Potting, J., Hekkert, M. P., Worrell, E., and Hanemaaijer, A. Circular Economy: Measuring Innovation in the Product Chain. PBL Netherlands Environmental Assessment Agency. 2017.

[21] Hunger, T., Arnold, M., and Ulber, M. Circular value chain blind spot – A scoping review of the 9R framework in consumption. *Journal of Cleaner Production*. 2024. 140853.

[22] PlasticsEurope. *The Circular Economy for Plastics – A European Analysis*. PlasticsEurope: Brussels. 2024.

[23] United States Environmental Protection Agency (EPA). *Advancing Sustainable Materials Management: Facts and Figures 2013. Assessing Trends in Material Generation, Recycling and Disposal in the United States*. Washington DC. 2015.

[24] Shin, D. *Generation and Disposition of Municipal Solid Waste (MSW) in the United States – National Survey*. M.Sc. Thesis, Columbia University. 2014.

[25] Santos, B. *Plastic packaging recycling rate breaks record in Germany*. Access Date: 16.02.2025. Available: *https://www.sustainableplastics.com/news/plastic-packaging-recycling-rate-breaks-record-germany*.

2

Plastics Recycling – Conservation of Valuable Resources

Plastics recycling is the term used for reprocessing postconsumer and preconsumer plastic waste (manufacturing scrap) into usable products. The idea behind recycling is to break down finished products into their component materials and then use those materials as feedstock to manufacture new products. Based on the plastic waste source, the recycling process and the finished product differ. In general, plastics can only be reused a limited number of times before they are too degraded for further use. Currently, all preconsumer plastic waste is fed back into the plastic production stream, but only a little portion of postconsumer plastic waste is reclaimed for its original use. However, every bit of plastic that is recycled reduces the need for new plastic feedstock and thus decreases the amount of resources and energy used for its production.

Recycling of plastics for use in creating new high-quality plastic products requires that the recycled materials are clean and consist of only a single type of plastic. In such cases, the recycled plastic substitutes for virgin plastic. The big challenge in recycling postconsumer plastics, especially those from mixed (co-mingled and/or single) stream collection, is that they are often contaminated. Recycling of mixed plastics is much more complicated. If the recycled plastics are contaminated and/or are a mixture of different types of plastic, the quality of the recycled plastic is lower; for example, the plastic may have lower strength. The challenge in managing the recycling of large quantities of a mixture of miscellaneous types of contaminated plastics needs to be considered using an integrated approach to source reduction, reuse, and recycling [1].

Plastics recycling is more complex than metal or glass recycling because of the many different types of plastic. Thus, recyclability and environmental compatibility need to be criteria considered at the beginning of the design process of plastic products instead of as an afterthought, particularly in many products where several kinds of plastic and sometimes non-plastic components are integrated. The separation, recovery, and purification of the plastic components in such a product require several steps, which consume additional energy. Unfortunately, the recycling rate, the amount of any type of plastic that is recycled in a period of time, is directly related to the price of virgin resins for that type of plastic, which is related to the price of oil (see Section 4.2.3). Low oil prices result in low costs for the virgin resins. In these times, recycled resins are too expensive to be used by comparison, and the recycling rates drop. Therefore, the goal of any sustainable growth in recycling should be the maximization of efficiency of energy utilization in every step of the process, from the initial production of plastic goods to the disposal or recovery of plastic wastes [2].

2.1 Plastics Recycling Methods

There are three common methods for plastics recycling: *mechanical recycling* (subdivided into primary and secondary recycling) and *chemical recycling* (tertiary recycling). The choice of the appropriate recycling method largely depends on the purity of the waste stream (Section 2.5), which can be contaminated with organic or inorganic substances, other polymers, or impurities. The molecular structure of the plastics as well as existing cross-links, such as in thermosets or rubbers, also influence the decision process [3, 4].

Figure 2.1 gives a schematic overview of the types of recycling methods, in terms of physical, chemical, and thermochemical processes.

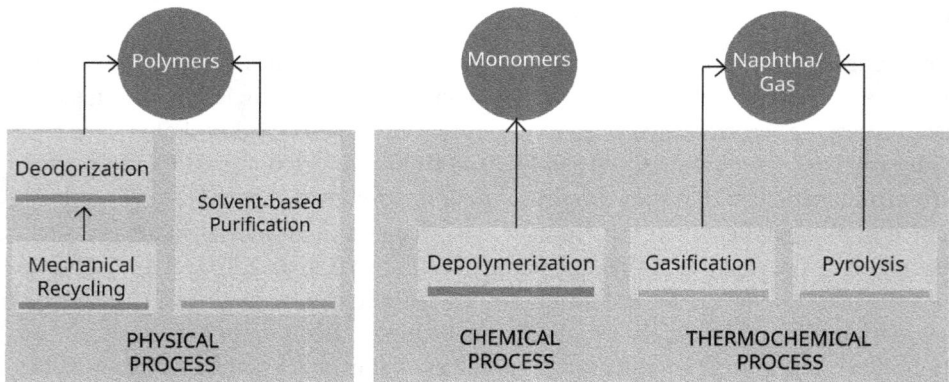

Figure 2.1 Schematic overview of the types of recycling methods

2.1.1 Mechanical Recycling

Among the recycling methods, mechanical recycling is the most desirable approach because of its low cost and high reliability. In general, mechanical recycling maintains the molecular structure of the polymer molecule. After grinding of the plastic waste material, the main processing step is remelting of the regrind material, which limits the use of mechanical recycling to thermoplastic polymers. Since remelting causes a degradation of the polymer chains, virgin material is often mixed with recycled material to reduce the effects of degradation on the product properties. The mixing leads to a dilution of the virgin material, which is described in Section 2.2.1.2 [5].

Mechanical recycling is divided into primary and secondary mechanical recycling, depending on whether the source of the waste is preconsumer or postconsumer, respectively. Preconsumer manufacturing scrap plastic is usually clean and of a single type or at least of a known composition and requires no further treatment, whereas postconsumer waste is highly contaminated and requires additional steps like collecting, sorting, and cleaning.

2.1.2 Chemical Recycling

Chemical recycling is used for *cross-linked polymers* that cannot be reprocessed mechanically or for thermoplastic polymers when mechanical recycling no longer yields sufficient quality due to contamination or material degradation. Through various chemical processes, polymer chains are broken down into *low molecular weight* compounds or, in some cases, their original monomers (feedstock). The recovered monomers can be repolymerized to produce virgin-quality plastics, while the low molecular weight compounds serve as feedstock for the petrochemical industry.

This field is evolving rapidly, driven by the need for higher-quality recyclates, improved process efficiency, and broader acceptance of chemically recycled materials in regulatory frameworks. Advances in catalysis, process optimization, and energy efficiency are expected to make chemical recycling more viable on an industrial scale. However, scalability, high energy consumption, and environmental trade-offs remain challenges. Different countries and organizations classify chemical recycling technologies using various frameworks, often focusing on distinctions between thermal and chemical methods as well as the nature and usability of the resulting products. The following classification is used in this book:

Solvent-based purification is a process in which polymers are dissolved in specific solvents to remove impurities such as colorants, additives, and other contaminants. Unlike chemical recycling methods that break down polymers into smaller molecular units, this technique preserves the original polymer chains, resulting in high-purity recyclates with properties similar to virgin materials [6]. Because of its ability to effectively remove both surface and internal contaminants, solvent-based purification produces cleaner and higher-grade recycled plastics than mechanical recycling [7].

While this method is promising, it is not a perpetual recycling solution. Each cycle introduces risks of residual contamination and potential polymer degradation due to solvent exposure, limiting the number of times a polymer can be reprocessed before it becomes unsuitable for further recycling [8]. Additionally, solvent recovery and removal are highly energy-intensive, making large-scale adoption challenging due to increased costs and environmental concerns. Despite these limitations, significant efforts are being made to commercialize and refine solvent-based purification. Additionally, solvent recovery and removal are highly energy-intensive, making large-scale adoption challenging due to increased costs and environmental concerns. Several companies have integrated this approach into their sustainability initiatives, particularly for polyethylene (PE) and polypropylene (PP), to enhance circularity and reduce reliance on virgin plastics.

Depolymerization, also referred to as decomposition or chemolysis, is the process of breaking polymers down into their original monomers or smaller molecules that can be repolymerized into new plastics. This method is particularly effective for *condensation polymers* such as polyethylene terephthalate (PET), polyurethanes (PU), polycarbonates (PC), and polyamides (PA) because their chemical structure allows them to be selectively broken down under controlled conditions. These monomers can then be purified and reintegrated into plastic production, allowing for true closed-loop recycling.

However, depolymerization is significantly more challenging for *addition polymers* like polyethylene (PE) and polypropylene (PP) due to their chemical inertness and strong carbon–carbon bonds. Traditional depolymerization processes for PE and PP, such as pyrolysis, require extreme temperatures and often produce non-selective hydrocarbon mixtures instead of recoverable monomers, limiting their viability for high-quality recycling.

Catalysts play a crucial role in improving the efficiency and selectivity of depolymerization. Without catalysts, depolymerization typically requires high temperatures and pressures, leading to increased energy consumption and operational costs. Research into *advanced catalysts*, *solvent-assisted depolymerization*, and *enzymatic approaches* is ongoing, with the goal of making chemical recycling more economically viable for industrial applications [9, 10].

Several depolymerization techniques exist, each with specific applications:

Glycolysis uses glycols to break down polyesters like PET into monomers such as ethylene glycol and terephthalic acid [11].

Hydrolysis employs water under high temperature and pressure to decompose polymers into their base monomers [12].

Methanolysis utilizes methanol to break down polyesters, particularly PET [13].

Enzymatic depolymerization leverages enzymes such as cutinases and PETases to break down polyester at mild temperatures, making it one of the most sustainable emerging methods [14].

Pyrolysis is a thermal decomposition process that heats plastics in an oxygen-free environment, breaking them down into hydrocarbon oils or gases. Depending on process conditions, pyrolysis can be adjusted to favor lighter or heavier hydrocarbons, which can serve as fuels or as feedstock for the production of new polymers. However, unless pyrolysis oil undergoes further refining into monomers, it does not fully close the plastic loop. The process has been criticized for its high energy consumption, CO_2 emissions, and reliance on fossil-fuel-based refining infrastructure, raising concerns about its long-term sustainability [15].

Gasification, in contrast, operates at higher temperatures and introduces a controlled amount of oxygen, converting plastics into syngas, a mixture of hydrogen and carbon monoxide. Syngas can be used for chemical synthesis, power generation, or hydrogen production, making gasification the most versatile chemical recycling method. It is capable of processing all types of plastic waste, including highly contaminated mixed plastics and hazardous waste from specialized collection systems [16]. However, most syngas today is used for energy recovery rather than material circularity, leading to debate over whether gasification should be classified as a true recycling method.

Both pyrolysis and gasification offer **important pathways for mixed plastic waste**, particularly in cases where mechanical and solvent-based methods are no longer viable. However, **optimizing reaction conditions, improving catalyst efficiency**, and **reducing environmental impact** remain key challenges for their large-scale deployment. Chemical recycling encompasses a range of technologies, each with distinct processes, end products, advantages, and limitations. The choice of method depends on factors such as polymer type, contamination level, and desired product quality. Table 2.1 provides a comparative overview of key chemical recycling technologies, outlining their process mechanisms, outputs, benefits, and challenges.

The environmental impacts and scalability of chemical recycling remain topics of debate. While these methods offer solutions for plastic waste that cannot be mechanically recycled, they also come with high energy demands, emissions, and by-product management challenges. Life cycle analysis (LCA) plays a crucial role evaluating the true sustainability of these processes, but current LCAs face limitations due to inconsistent methodologies, data gaps, and difficulties in emissions accounting. Regulatory frameworks further complicate the adoption of chemical recycling. Different countries define and categorize chemical recycling differently, impacting whether chemically recycled plastics count toward recycled content targets and circular economy goals. Policy measures such as CO_2 penalties, extended producer responsibility (ESP) schemes, and stricter packaging regulations will play a critical role in shaping the future of chemical recycling. Ultimately, improving plastic design for mechanical recyclability remains the most effective strategy for reducing reliance on energy-intensive downstream processes like pyrolysis and gasification

Table 2.1 Comparison of Chemical Recycling Technologies

Technology	Process	Products	Advantages	Limitations	Current Commercial Status
Solvent-based purification (physical recycling)	Selective dissolution of polymers in solvents, followed by filtration and polymer pre-cipitation	High-purity polymer granules	Preserves polymer structure, produces high-quality recyclates, adaptable to mixed plastic waste	High solvent recovery energy demand, limited scalability, solvent toxicity concerns	Emerging, small-scale commercial applications
Depoly-merization	Chemical breakdown of polymers into monomers	Terephthalic acid + ethylene glycol (PET), bisphenol A (PC), lactam (PA)	Produces virgin-quality monomers, supports true circular recy-cling	High energy and catalyst costs, limited to conden-sation poly-mers, challeng-ing for PE/PP	Commercial for PET, pilot-scale for other plastics
Pyrolysis	High-tem-perature breakdown (250–700 °C) in an inert atmosphere	Pyrolysis oil, naphtha, waxes, syngas	Processes a broad range of mixed plastics, including con-taminated waste	High energy consumption, requires exten-sive refining for reuse in plastics	Expanding commercial applications with major investments
Gasification	Very-high-temperature breakdown (up to 1500 °C) with limited oxygen	Syngas ($CO + H_2$), methanol, ammonia	Processes highly contam-inated and hazardous plastics, enables waste-to-chemical conversion	Mostly used for energy recovery rather than plastic-to-plastic recy-cling, requires significant infrastructure	Limited com-mercial use, mostly in waste-to-energy plants

2.2 Recycling Different Types of Plastic Waste

As mentioned before, plastic waste can be divided into *preconsumer waste* (manufac-turing scrap) and *postconsumer waste* (recovered waste). These different plastic waste types are recycled differently.

2.2.1 Preconsumer Waste

2.2.1.1 Manufacturing Scrap

Preconsumer waste, such as runners, gates, sprues, and trimming, is normally recycled using primary mechanical recycling. It is ground and remelted in-house.

2.2.1.2 Dilution Effect

Manufacturing scrap is often mixed into virgin material to reduce material cost while at the same time minimizing the effects of degradation on part performance. Depending on the mixing ratio, either the virgin material is diluted with regrind or the regrind is refreshed with virgin material. By using a constant mixing ratio during continuous processing, the regrind waste itself is diluted by material that has been reprocessed once, twice, three times, etc. The composition of a material with a proportion of recyclate q after n processing cycles can be calculated using Equation 2.1.

$$\sum_{i=1}^{n} q^{n-i} (1 - q) = 1 \tag{2.1}$$

For small proportions of recyclate, the regrind material contains only minimal amounts of material that has passed through a large number of processing cycles and therefore is highly degraded.

Figure 2.2 shows the composition of material with different mixing ratios of recycled and virgin material. The first column shows 30% recycled and 70% virgin material. Under these conditions, the regrind material contains less than 0.8% of material that has been reprocessed five times or more. Seventy percent of the material is virgin material, 21% has been processed once, 6.3% twice, and 1.9% three times. As proportions of material smaller than 1% do not have a significant influence on the material properties and can be neglected [17], the properties will be dominated by fractions that have been processed four times or less. Thus, it can be concluded that the properties of a material with small amounts of recyclate will not fall below a certain level [18].

However, regrind material with high proportions of recyclate contains significant amounts of highly degraded material, as can be seen in the right bar in Figure 2.2, in which 70% of the regrind is recycled and 30% is virgin material. This regrind material contains 5.0% material that has been reprocessed five times, as well as 30% that is virgin material, 21% that has been processed once, 14.7% twice, 10.3% three times, and 7.2% four times. After nine processing cycles, the material still contains 1.2% of the initial material. Although this mix contains significant portions of highly degraded material, after 10 reprocessing cycles the material reaches a steady state in which performance properties are not affected anymore by further processing. Therefore, this mixing ratio is used quite frequently for packaging products, e. g., PET containers.

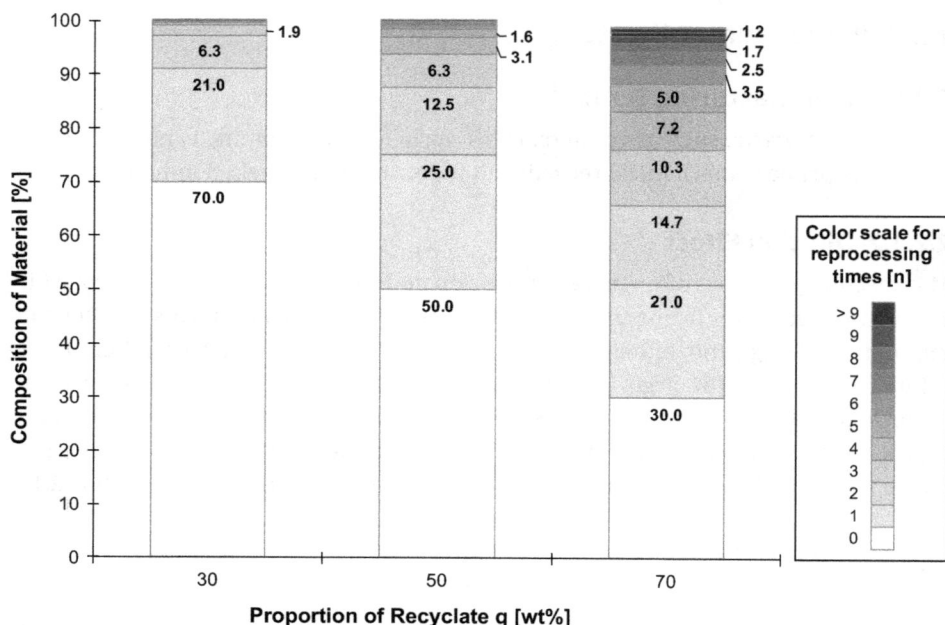

Figure 2.2 Composition of recycled plastic material after **n** reprocessing steps for 30%, 50%, and 70% recycled material

2.2.2 Postconsumer Waste

Consumer plastics are largely made from six different polymer resins, which are indicated by a number, or *resin code*, from 1 to 7 molded or embossed onto the surface of the plastic product. The number 7 indicates any polymer other than those numbered 1 to 6. Table 2.2 lists the polymer resins, their resin codes, and the general applications for virgin and recycled plastics made from these resins. The percentages of the different types of postconsumer plastic waste in municipal solid waste (MSW) in the United States in 2017 are given in Table 2.2 [19].

The chemical composition and function of each resin controls where the resin can be recycled as well as the recycling rate. The latter is attributed to the difficulty of separating mixed plastic during the recycling process. For example, PET, or resin code 1, only accounts for 14.39% of the total plastic waste but it has the highest recycling rate of all resins. Because of its widespread use in transparent drinking bottles, PET is easy to identify and sort by transmission detectors.

Table 2.2 Plastic Types and Products from Virgin and Recycled Materials

Resin Symbol and Plastic Type	Products Created from Virgin Plastics	Products Created from Recycled Plastics
01 PET Polyethylene terephthalate	Bottles for water, soft drinks, salad dressing, peanut butter, and vegetable oil	Egg cartons, carpet, and fibers and fabric for T-shirts, fleeces, tote bags, shoes, etc.
02 HDPE High-density polyethylene	Milk and juice cartons, detergent containers, shower gel bottles, and shipping containers	Toys, pails, drums, traffic barrier cones, fencing, and trash cans
03 PVC Polyvinyl chloride	Packaging materials, plastic pipes, decking, wire and cable products, blood bags, and medical tubing	Shoe soles, construction material, and boating and docking bumpers
04 LDPE Low-density polyethylene	Disposable diaper liners, cable sheathing, shrink-wrap, and film	Timbers, trash can liners, shopping envelopes, lumber, and floor tiles
05 PP Polypropylene	Medicine bottles, drinking straws, yogurt containers, butter and margarine tubs, automotive parts, and carpeting	Signal lights, bicycle racks, trays, battery cables, and ice scrapers
06 PS Polystyrene	Egg cartons, cups, food containers, plastic forks, and foam packaging	Egg cartons, foam packing, and light-switch plates
07 O All other resins or mixtures of resins	Mixed plastics or multilayer plastics packaging	–

Some resins are not compatible with others, because their molecular structures repel each other if mixed. This leads to deterioration of the mechanical performance of plastic products made from them if they are not engineered properly. Most plastics have additives incorporated to achieve certain additional properties such as flame retardancy, flexibility, or resistance to ultraviolet (UV) damage. This makes it nearly impossible to obtain a homogeneous plastic mixture with uniform behavior. Therefore, it is important that the sorting process is well regulated to ensure the integrity and overall performance of recycled plastic products.

Depending on their properties, different plastics are used for different applications. Currently, packaging, consumer and institutional products, and building and construction materials are the top three uses for plastics. The share of U.S. plastics demand by use in 2017 is shown in Figure 2.3 These different applications are again represented in the plastic waste [20].

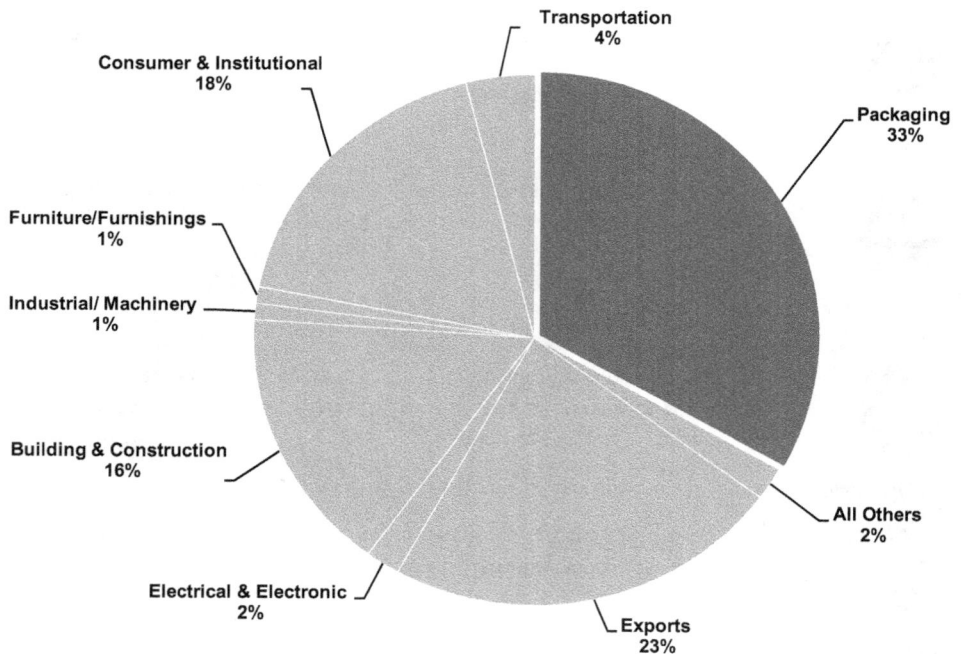

Figure 2.3 United States plastics demand by use in 2017 [19]

2.2.2.1 Packaging Plastic Waste

As a result of the demand for plastics, a large share of MSW plastics consists of packaging items in which high-density polyethylene (HDPE, 17.16%), low-density polyethylene (LDPE, 22.94%), and polypropylene (PP, 22.76%) together account for about 63% of the waste (see Table 2.1). The continuing increase in the use of disposable packaging has led to increasing amounts of plastics ending up in the waste stream. In Europe, where the proportion of plastic waste in the waste stream is similar to that in the United States, packaging waste amounts to nearly two-thirds of all the plastic waste, as shown in Figure 7.6 [21]. The difference in the demand for packaging (34%) and the generation of packaging waste (63%) is based on the different service life of the products. About 60% of all plastic products were designed for a long service life (years), while about 40% were designed for a shorter service life or even for a single use.

Due to their similar density, packaging wastes, especially LDPE, HDPE, and PP, are difficult to separate (Section 2.3.2). Unfortunately, waste management strategies are not developing at the same rate as the increasing levels of plastic waste. In 2017, only 3 million tons (9.23%) out of the total 32.5 million tons of plastic waste in America's MSW stream were recycled [19, 22]. The recovered plastics were mainly PET and HDPE bottles. The main polymers used for packaging applications are summarized in Table 2.3 [21].

Table 2.3 Most Common Polymers Used in Packaging Applications

Application	Most Common Polymers Used
Bottles, flasks	PET (66%), HDPE (28%), PP (3%), LDPE (0.4%), PVC (3%)
Closure items, bottle caps	PP (73%), HDPE (20%), LDPE (5%), PVC (2%)
Films	LDPE (76%), PP (20%), PVC (3%), PET (1%)
Bags, sacks	LDPE (61%), HDPE (31%), PP (8%)
Jars, boxes, tubs	PP (73%), HDPE (20%), LDPE (5%), PVC (2%)

Abbreviations: HDPE, high-density polyethylene; LDPE, low-density polyethylene; PET, polyethylene terephthalate; PP, polypropylene; PVC, polyvinyl chloride

2.2.2.2 Building and Construction Plastic Waste

In the building and construction industry, plastics play an important role due to their durability, aesthetics, easy handling, and high performance. Normally, they are designed to be durable for 30 to 40 years before disposal. They are used, for example, in pipework, insulation, wall coverings and flooring, interior fittings, and window frames. Common plastics used in construction include HDPE, polyvinyl chloride (PVC), and polyurethane (PU) (see Table 2.4) [21].

Table 2.4 Most Common Polymers Used in Building and Construction Applications

Application	Most Common Polymers Used
Pipes and ducts	PVC, PP, HDPE, LDPE, ABS
Insulation	PU, EPS, XPS
Windows and other frames, flooring, and wall coverings	PVC
Lining	PE, PVC
Interior fittings	PS, PMMA, PC, POM, PA

Abbreviations: ABS, acrylonitrile butadiene styrene; EPS, expanded polystyrene; HDPE, high-density polyethylene; LDPE, low-density polyethylene; PA, polyamide; PC, polycarbonate; PE, polyethylene; PMMA, polymethyl methacrylate; POM, polyoxymethylene; PP, polypropylene; PS, polystyrene; PU, polyurethane; PVC, polyvinyl chloride; XPS, extruded polystyrene

Rigid PU foam is known for its high thermal resistance, which promotes temperature maintenance. It is also popular because it is lightweight, chemically resistant, and flame retardant. Moreover, acrylonitrile butadiene styrene (ABS) and polycarbonate (PC) are used in the construction industry as well.

The variety of material grades and properties used in the building and construction industry leads to a small recycling rate for these materials [23]. Nevertheless, the building and construction industry can be counted as a secondary market for recycled plastic. There it can be used in many applications including as a filler, in packaging, in landscaping for walkways, bridges, fences, and signs, and in traffic management and industrial strapping products [24].

2.2.2.3 Automotive Plastic Waste

In automotive applications, plastics are used for a wide variety of parts and functions, mainly because of their low weight and cost. Furthermore, they have advantages due to their impact and corrosion resistance. The largest share of plastics used in vehicles is for the passenger cell, followed by the body. PP, PE, PU, and PVC are the most common plastics by volume used in a typical car. Table 2.5 lists the common plastics used in various automotive parts.

Table 2.5 Most Common Polymers Used in Automotive Applications

Application	Most Common Polymers Used
Bumper	PP, ABS, PC/PBT
Seats	PU, PP, PVC, ABS, PA
Dashboard	PP, ABS, PPE, PC

Application	Most Common Polymers Used
Fuel system	HDPE, POM, PA, PP, PBT
Body	PP, PPE
Interior trim	ABS, PP, PBT, POM, PP
Lighting	PC, PBT, ABS, PMMA

Abbreviations: ABS, acrylonitrile butadiene styrene; HDPE, high-density polyethylene; PA, polyamide; PBT, polybutylene terephthalate; PC, polycarbonate; PMMA, polymethyl methacrylate; POM, polyoxymethylene; PP, polypropylene; PPE, polyphenylene ether; PU, polyurethane; PVC,

2.2.2.4 Agricultural Plastic Waste

Plastics are everywhere on a typical farm and substitute for traditional materials due to their low price and light weight. Nowadays, hay bales are often wrapped in plastic or grain is stored in plastic bags instead of silos. Furthermore, plastic has become essential for widespread applications in modern farm operations. For example, large plastic sheets are used for everything from heating the soil and suppressing weeds to roofing greenhouses. A lot of agricultural products are plastic films, which is reflected in the types of polymers used in agricultural applications as shown in Table 2.6. Currently only 10% of agricultural plastic waste is recycled [21, 25, 26, 27].

Table 2.6 Most Common Polymers Used in Agricultural Applications

Application	Most Common Polymers Used
Bale bags, seed bags	LDPE, LLDPE, PP
Greenhouse covers, silo covers, mulch film	LDPE, LLDPE
Nets and mesh	LDPE, HDPE
Rope, strings	PP
Pipes and fittings	PVC, LDPE
Pesticide containers, nursery pots	HDPE, PS, PP

Abbreviations: HDPE, high-density polyethylene; LDPE, low-density polyethylene; LLDPE, linear low-density polyethylene; PP, polypropylene; PS, polystyrene; PVC, polyvinyl chloride

2.2.2.5 Waste from Electrical and Electronic Equipment (WEEE)

In general, the plastic waste from electrical and electronic equipment (WEEE) and in particular the metals in them are recovered by metal recyclers. The remaining plastics and nonmetals are known as electronics shredder residue (ESR). The principal polymers found in WEEE include ABS, high-impact polystyrene (HIPS), PP, polyamide

(PA), polycarbonate (PC), blends of PC with ABS (PC/ABS), blends of polyphenylene ester (PPE) and HIPS (PPE/HIPS), and some others as shown in Table 2.7. So far, only half of a typical ESR mixture is recovered and the rest goes into landfill [21, 28, 50].

Table 2.7 Most Common Polymers Used in Electrical and Electronics Equipment

Application	Most Common Polymers Used
Printers/faxes	PS, HIPS, SAN, ABS, PP
Telecommunications equipment	ABS, PC/ABS, HIPS, POM
Televisions	PPE/HIPS, PC/ABS, PET
Monitors	PC/ABS, ABS, HIPS
Computers	ABS, PC/ABS, HIPS
Refrigeration	PS, ABS, PU, PVC
Dishwashers	PP, PS, ABS, PVC

Abbreviations: ABS, acrylonitrile butadiene styrene; HIPS, high-impact polystyrene; PC, polycarbonate; PET, polyethylene terephthalate; POM, polyoxymethylene; PP, polypropylene; PPE, polyphenylene ether; PS, polystyrene; PU, polyurethane; PVC, polyvinyl chloride

2.3 Sorting Processes for Plastic Waste

After postconsumer plastic waste is collected, it is transported to material recovery factories (MRF) and, in a first step, sorted by plastic type. Depending on the type of plastic, sorting methods vary. Sorting currently is the step in the recycling process of postconsumer waste that has the largest impact on integrity. Sorting can be done manually or it can be automated. In the following discussion, the main methods used in the MRF in the United States for the different plastics are identified.

2.3.1 Manual Sorting

A simple sorting method is *manual sorting*. It involves visual identification of the plastic type by operators using the resin identification code, shape, color, appearance, and trademark of the plastic. It is very labor intensive and has a possibility of human error. Furthermore, it is difficult to manually differentiate between the resin types by just visual means.

2.3.2 Automated Sorting

2.3.2.1 Float-and-Sink Sorting

The *float-and-sink* process, or sorting by flotation, in which plastics are sorted by density is one of the most common *automated sorting* processes. The washed and chipped plastics are sent into tubs of water and the pieces that float or sink are separated. This process is fast, inexpensive, and can be considered as a first stage washing of plastic waste. However, as pointed out earlier, most plastics are very similar in density and thus cannot be separated using this process. An overview of the densities of various polymers is given in Table 2.8 [17, 29, 30].

Table 2.8 Polymer Density Ranges (densities of < 1 g/cm^3 will float)

Polymer	Density Range [g/cm^3]
Polyethylene terephthalate (PET)	1.330–1.400
High-density polyethylene (HDPE)	0.956–0.980
Polyvinyl chloride (PVC)	1.304–1.388
Low-density polyethylene (LDPE)	0.910–0.955
Polypropylene (PP)	0.861–0.925
Polystyrene (PS)	1.050–1.220

2.3.2.2 Froth-Flotation Sorting

The *froth-flotation* process works similarly to the float-and-sink process. In froth-flotation, the materials to be separated are first treated with a surfactant and then suspended in water. Plastics that would normally sink in water are suspended in the water-surfactant mixture. Then, air is pumped into the system. The air bubbles adhere to some plastic pieces based on their resin type but others are not affected by the air bubbles and so sink to the bottom. The key advantage of this method is to be able to separate PET from PVC.

2.3.2.3 Near-Infrared Sorting

Another promising technology suitable for high-speed sorting machines is *near-infrared (NIR) sorting*, which uses a well-established technique for characterizing the type of plastic, infrared light transmission. Plastics absorb light of specific wavelengths unique to their chemical composition, allowing for the identification and separation of different types of plastic.

This technology is mostly used for the automated identification of PET and HDPE bottles. However, this sorting method has limitations when it comes to dark-colored plastics, multilayer products, thin films, surface contaminated materials, materials with high or no reflectance as well as adhesives, residues, and plastics with additives [31, 32].

2.3.2.4 Mid-Infrared Sorting

Mid-infrared (MIR) sorting has emerged as a solution to the limitations of near-infrared (NIR) technology for dark-colored plastics, as these pigments do not fully absorb in the MIR range. However, its adoption remains slow due to higher investment costs and increased energy consumption. Additionally, the heat generated by the radiation source presents a fire hazard, restricting its use primarily to fundamental applications such as sorting HDPE, PP, PET, PVC, and PS.

2.3.2.5 AI-Based Sorting

AI is revolutionizing plastic sorting by addressing the limitations of conventional technologies like near-infrared (NIR) and density-based separation. AI-based systems use artificial neural networks (ANNs) to analyze and classify materials more accurately. These networks consist of interconnected nodes that process and weight information across multiple layers, continuously improving sorting precision. Unlike traditional methods that rely on spectral signatures, AI integrates optical recognition and machine learning. Cameras detect size, shape, and other optical characteristics, allowing differentiation beyond material composition. By training on large datasets, AI algorithms enhance their ability to sort mixed plastic waste, even when conventional techniques struggle with contamination, complex structures, or multi-material packaging. However, these systems are still in their early stages and require extensive training to reach their full potential.

AI-assisted sorting is particularly useful where existing methods fall short, such as separating silicone cartridges and multilayer plastics, distinguishing food-grade materials, sorting HDPE and PP from mixed waste, and improving PET recovery despite contamination. By distinguishing plastic grades within the same material group, AI increases the purity and yield of recyclates. Its self-learning capability enables continuous optimization, improving waste separation and generating new product streams. Despite its potential, AI sorting faces challenges, including high data processing demands and the need for large training datasets. AI models must be trained extensively to recognize diverse plastic variations, and their effectiveness depends on data quality. Additionally, high initial investment costs remain a barrier to widespread adoption in the recycling industry.

2.3.2.6 Laser-Based Identification

Laser-based systems identify plastics by directing a laser beam onto the material, analyzing spectral and spatial characteristics for precise identification and positioning, making them well-suited for high-speed sorting lines.

Some advanced laser systems can also determine material properties such as absorption coefficient, thermal conductivity, heat capacity, and surface temperature distribution, further aiding in plastic classification. Primarily used for sorting plastic flakes, these systems are effective for dark-colored materials. However, due to their complexity and high investment costs, their application has so far been largely limited to waste electrical and electronic equipment (WEEE) and automotive plastic recycling.

2.3.2.7 Electrostatic Sorting

Electrostatic sorting uses the triboelectric effect to separate plastics based on their electrical charge, rather than density or color. This method is effective for 6–12 mm plastic flakes, including black plastics, which are difficult to detect with optical sorting.

This method is applied in recycling streams where conventional techniques struggle with material differentiation. Key applications include PVC separation from window profiles, ABS and PS recovery from electronic waste, and PET/PVC sorting in beverage bottle recycling. It is also used for separating PE and PP in packaging waste and in the automotive industry for recovering ABS and PMMA from vehicle components.

2.3.2.8 X-Ray Fluorescence

Some plastics, like PET and PVC, have similar densities, and so sorting them must be based on another property in which they differ significantly. The molecular structure of PVC includes chlorine atoms, unlike PET, which does not. *X-ray fluorescence* generates a spectral fingerprint based on a plastic's chemical composition, and, from this, it can be sorted into its resin type. For example, due to the detection of the chlorine atoms in PVC, it can be used to separate PVC from other plastics [33].

2.3.2.9 Marker-Based Sorting Systems

Marker-based sorting systems involve embedding detectable markers either within the plastic itself or on the product surface. These markers allow for precise identification of resin type, color, additives, and even previous contents. While early marker systems, such as those developed by Continental Container Corporation and Eastman Chemical Company, demonstrated the potential for advanced sorting, their high implementation costs and limited industry adoption kept them at the proof-of-concept stage. One newer approach embeds fluorescent particles into or onto plastic products during manufacturing, which are later activated and detected during sorting. This technology achieves high purity and recovery rates and can be combined with other sorting methods, though its success depends on the availability of appropriate recycling infrastructure. This is most applicable in closed-loop recycling systems, where material flows remain controlled.

Digital watermarks represent another innovative approach, embedding an invisible code within the printed design of packaging. These watermarks can be recognized by cameras during sorting, enabling highly specific material identification. However, this technology is still in the early stages of development and has yet to achieve widespread market adoption.

2.4 Plastic Degradation Mechanisms

In general, polymers are stabilized only for processing and a first lifetime of use and not for reprocessing or further lifetimes of use. This is particularly true for plastic film packaging material made out of polyolefins, such as PP and PE, which usually has a short lifetime.

Degradation of a polymer results from exposure to factors like heat, mechanical stress, oxygen, and UV light. For example, polyethylene films used in commercial packaging suffer from oxidative and photo-oxidative degradation due both to processing and to use. Reprocessing can induce mechanical, thermal, and thermal oxidative degradation.

The degradation process leads to changes in the polymer's structure, typically characterized by fragmentation, or scission, of the macromolecular chain, which results in changes in the properties of the polymer. Radical chain reactions, such as formation of hydroperoxides and cross-linking, also occur during degradation and also result in changes in the polymer's structure. During processing, both chain scission and cross-linking reactions take place, which lead to increased fractions of lower molecular weight polymer chains and increased chain branching, respectively. Thus, fragmentation, self-termination, and chain-transfer reactions of the polymer radicals are occurring during degradation [34]. Polyolefins often undergo oxidative degradation.

During processing and lifetime use, the material is exposed to heat, mechanical stresses, oxygen, and UV radiation. Depending on the polymer structure and the processing conditions, these stresses lead to more or less degradation. In general, the effects of the degradation processes can be categorized as:

- Change in molecular weight and molecular weight distribution, leading to a change in viscosity

- Formation of cross-links and branched chains

- Formation of oxygenated compounds and unsaturated compounds

A knowledge of the degradation kinetics of polymers is important for designing the recycling process and the processes for creating new products out of recycled material. During the recycling process, the *melt viscosity* and *flow behavior* of materials change, leading to a change in the processing settings and sometimes difficulty in reprocessing. The recycled products can further have inferior mechanical properties compared to virgin material. The thermal stability is not affected for most polymers.

The mechanism of degradation, which can be classified as *mechanical degradation*, *thermal degradation*, *thermal oxidative degradation*, or *photo degradation*, depends on the method used for polymerization. Furthermore, the degree and type of degradation depend on the processing conditions and on the nature of the polymer.

2.4.1 Mechanical Degradation

The carbon–carbon chemical bonds of the polymer backbone break when shear and tensile stresses (mechanical stresses), induced by shearing and stretching in the extrusion process, exceed the intramolecular bonding forces leading to mechanical degradation. The mechanical stresses increase with increasing chain length and decreasing distance to the center of the chain. As a result, chain scission increases for longer polymer chains and is most likely to occur in the chain center. In addition, a reduction in temperature decreases the flexibility of the chain segments, which also leads to an increase of these mechanical stresses [35, 36].

2.4.2 Thermal Degradation

Thermal degradation is induced by heat. Heating of a polymer results in an increase of its internal energy causing the following effects: First, the rate of rotation of any freely moving group in the polymer increases, weakening the intermolecular forces. Second, the vibrational energy of the polymer bonds increases, leading to bond breaking along the chain according to a statistical pattern and the formation of two radicals for each end of the broken bond. Finally, the mobility of absorbed species increases, enabling their migration through the polymer and hence their reaction with energetic sites [37, 38].

2.4.3 Thermal Oxidative Degradation

Thermal oxidative degradation of polymers is caused by autoxidation. The reaction follows the same steps known for polymerization: initiation, propagation, and termination. Depending on the dominating reactions during autoxidation, either a decrease or an increase of the average molecular weight of the polymer can occur. While β-scission and fragmentation cause a decrease of the molecular weight, an increase can be observed for recombination [39, 40, 41].

2.4.4 Effect of Degradation on Processing and Service-Life Properties

It is important to analyze the effect of degradation on the processing parameters as well as the service-life properties of the desired product. Based on the product requirements, the number of reprocessing events has to be limited or the amount of virgin material mixed into the recyclate has to be adjusted to reduce the dilution effect. In the following, data for these property changes is presented for some common unfilled and filled plastics.

2.4.4.1 Unfilled Plastics

As already mentioned, PET is the most widely recycled polymer due to the ease of its separation in the recycling stream. The degradation mechanism of PET is chain scission, which can be seen from the reduction of molecular weight from virgin PET to recycled PET over three extrusion cycles (Figure 2.4 *left*). Even though the chain length is reduced significantly, the mechanical properties are only slightly affected as can be seen in Figure 2.4 *middle* and *right*. Both tensile strength and impact strength show a small reduction over the three extrusion cycles.

In industrial practice, virgin PET is often used to refresh the recyclate. The mixing ratio and resulting dilution (Figure 2.2) depends on the desired product properties. Figure 2.5 shows the change of the molecular weight and impact strength for different mixing ratios. The graph in Figure 2.4 *left* shows that the molecular weight of 100% recycled PET (same as reprocessing step 2 in Figure 2.4) is about 20% lower than virgin PET (same as reprocessing step 1 in Figure 2.4). The blended materials follow the linear trend based on the mixing ratio of virgin to recycled PET. Both data sets show the feasibility of reprocessing and the ability of refreshing with virgin material to slow down the degradation process significantly.

A decrease in molecular weight leads to a reduction in melt viscosity, which is a measure of the resistance to flow, and thus affects the processing behavior. Therefore, the process settings have to be adjusted accordingly. For example, the injection pressure can be lower. However, if the mixing ratio is kept constant, the material is in a steady state and adjustments are very minimal. Nevertheless, in certain cases the processibility of materials is affected so strongly that the material cannot be processed with the same process. The stretch-blow molding process is an example of such a sensitive process.

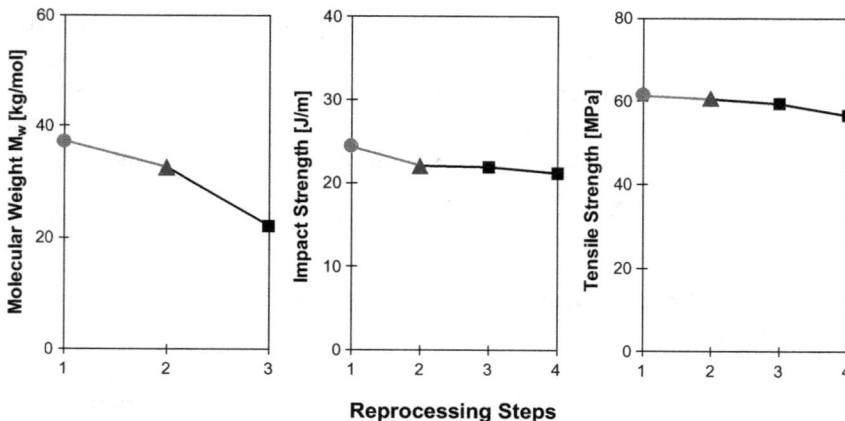

Figure 2.4 Weight-averaged molecular weight and mechanical properties of virgin polyethylene terephthalate (PET) and recycled PET as a function of the number of reprocessing steps (extrusion) [42]

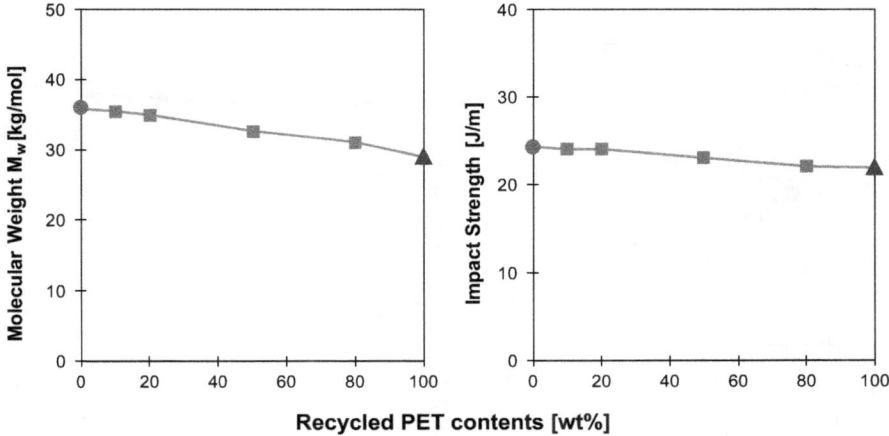

Figure 2.5 Weight-averaged molecular weight and impact strength of virgin PET (polyethylene terephthalate)/recycled PET blends during the first extrusion step [42]

High-impact polystyrene (HIPS) is among the promising materials for recycling since it shows only small variations in melt viscosity. Figure 2.6 shows the change in viscosity measured as melt flow index (MFI), which indicates how much material in grams flows through a small capillary within 10 min. The larger the MFI values, the lower the viscosity. The MFI increases slightly with an increasing number of reprocessing steps. The increase in MFI during this continued reprocessing without refreshing is caused by chain scission degradation due to thermal and mechanical degradation.

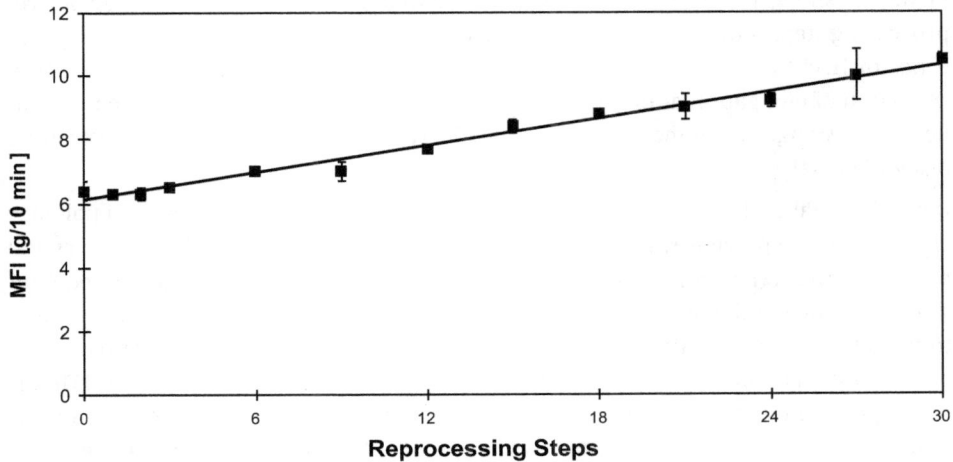

Figure 2.6 Melt flow index (MFI) of high-impact polystyrene (HIPS) as a function of the number of reprocessing steps (injection molding) [43]

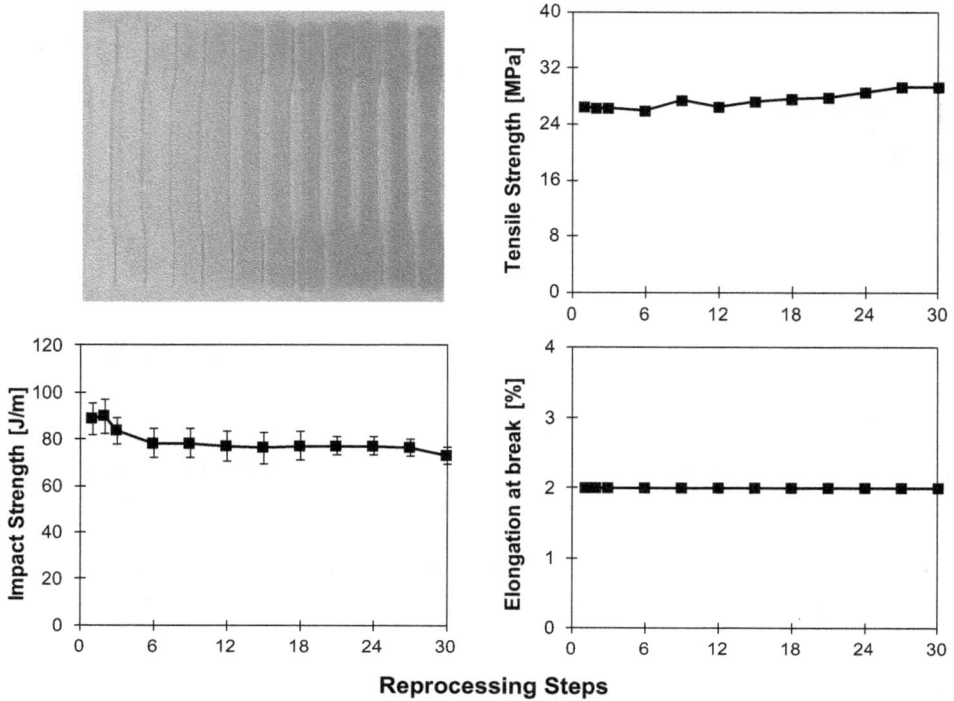

Figure 2.7 Discoloration and mechanical properties of high-impact polystyrene (HIPS) as a function of the number of reprocessing steps (injection molding) [43]

Figure 2.7 shows the discoloration of HIPS due to degradation during the various reprocessing steps. Furthermore, it illustrates the resulting changes in the tensile properties of HIPS. The degradation leads to an increase of tensile stress at break and a reduction of elongation at break over multiple reprocessing steps. Thus, the material behavior changes from more ductile to brittle, which needs to be known for product design [43, 44].

Instead of chain scission, degradation can lead to uncontrolled cross-linking of polymer chains for some polymers. This can be observed in the data of LDPE obtained over 100 extrusion cycles (Figure 2.8). Here, the complex viscosity of LDPE *(left)* increased with the increasing number of extrusion cycles. The cross-linking in LDPE chains occurs due to the formation and reaction of carbon radicals. The same trend was observed in the decreasing MFI values *(right)*. It is important to note that the reprocessing of LDPE is only significantly affected after the 40th extrusion cycle. Using a refreshing rate of less than 10% of virgin material results in a fraction of < 1% of 40 times processed material according to Equation 3.1 and thus has no practical relevance in this case [5].

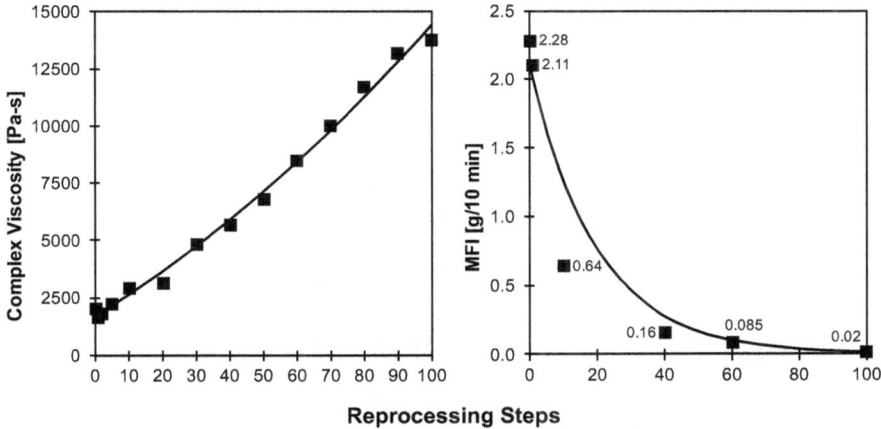

Figure 2.8 Complex viscosity and melt flow index (MFI) of low-density polyethylene (LDPE) during 100 reprocessing steps (extrusion) [5]

In addition to synthetic polymers, more and more biopolymers, which are made from renewable resources, are being recycled. One major market for biopolymers is plastic packaging. Polylactic acid (PLA), which is biodegradable under very specific conditions, is such a polymer. Since the conditions for chemical recycling can only be achieved in specific recycling facilities, the mechanical recycling of PLA is of interest as well. Figure 2.9 shows that multiple injection molding cycles cause degradation, which can be seen in the significant change in MFI. However, the impact strength of PLA remained unchanged even after six cycles, which indicates good recyclability in regards to the service-life properties of the plastic.

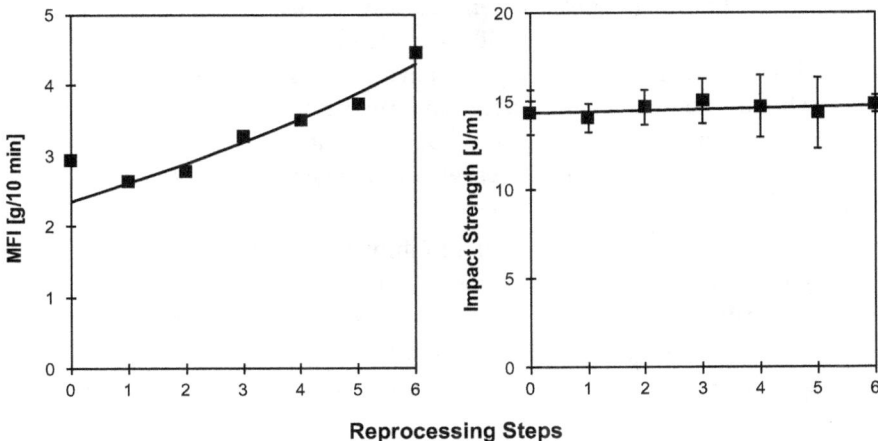

Figure 2.9 Melt flow index (MFI) and impact strength of polylactic acid (PLA) as a function of the number of reprocessing steps (injection molding) [45]

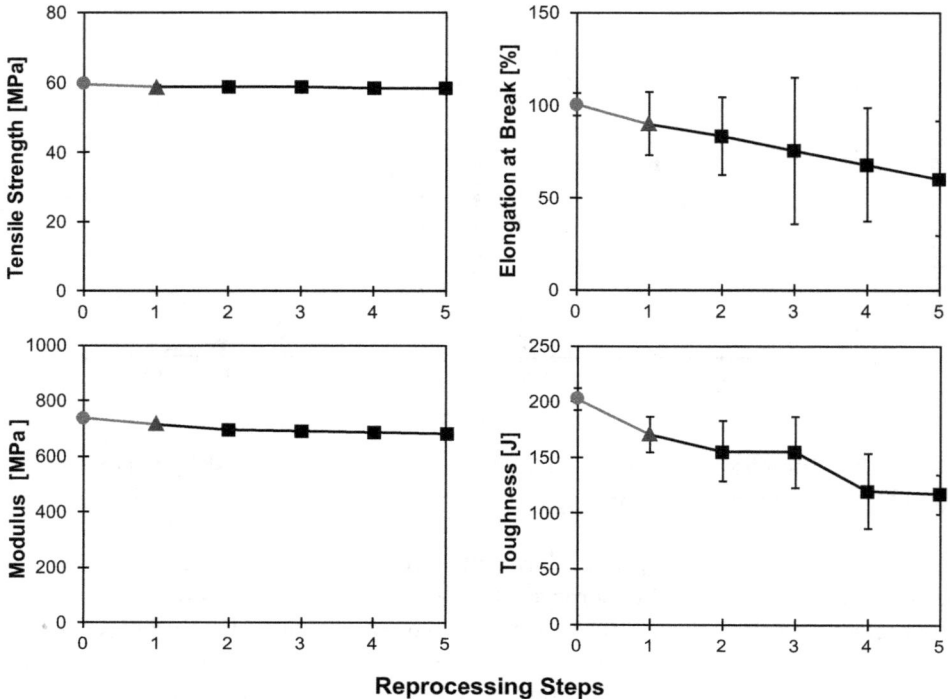

Figure 2.10 Mechanical properties of polycarbonate (PC) as a function of the number of reprocessing steps (injection molding) [46]

Polycarbonate (PC) is another material with a high potential for recycling. It is identified as plastic number 7 and mostly found in electronic waste. Recycling of PC is important to prevent the leaching of bisphenol A in PC into landfills. Figure 2.10 illustrates the effect of repeated injection molding cycles on the mechanical properties of PC. Tensile strength and modulus remain unchanged after five cycles. In contrast, the elongation at break and the toughness decrease with the increasing number of reprocessing cycles. The toughness of PC shows an approximately 42% reduction from the virgin resin to the final reprocessing steps, whereas the elongation at break shows a 37% reduction [46].

Again, comparing the property changes after multiple processing steps of 100% of the material (Figure 2.10) with the change in refreshed material reveals almost unchanged behavior during refreshing. Figure 2.11 shows the same mechanical properties, but for different mixing ratios of virgin and recycled PC. It can be seen that the material follows the linear trend depending on the mixing ratio, similar to the observations with PET.

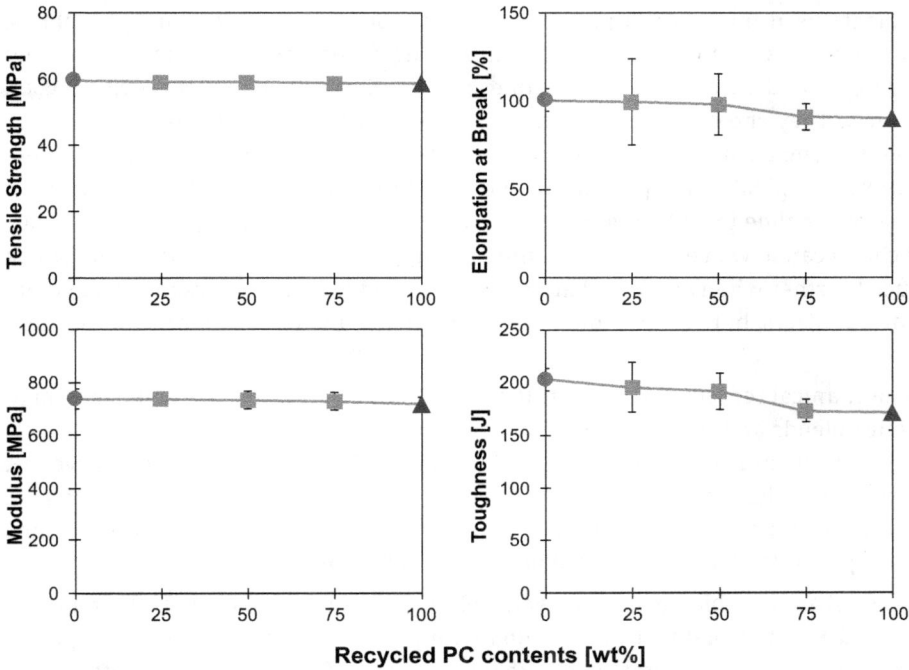

Figure 2.11 Mechanical properties of virgin polycarbonate (PC)/recycled PC blends as a function of the percentage of recycled PC, in the first reprocessing step [46]

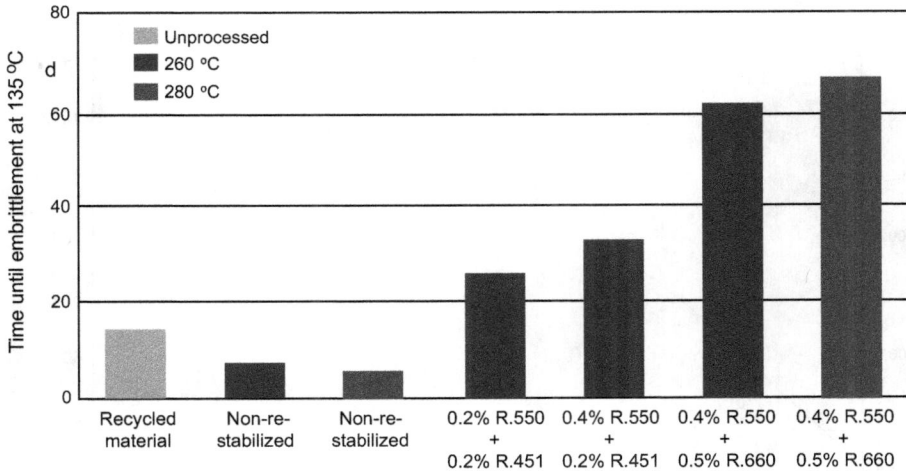

Figure 2.12 Aging stability of a CaCO₃-filled, recycled PP material from garden chairs at different processing temperatures (R. 451 = Recyclostab 451, R. 550 = Recyclossorb 550, R. 660 = Recycloblend 660) [61]

Stabilizers used for recycled plastics are mostly phenolic antioxidants, phosphites, or phosphonites combined with co-stabilizers, hindered amines, sterically hindered piperidines, and UV absorbers. Pfaendner [61] and Pospisil et al. [62] provide a good overview. They show in one study that a mixture of two additives, one used to improve the long-term stability of PP and the other a UV stabilizer with additional enhancement capability for processing and long-term stability of HDPE, significantly increases the time to embrittlement of filled PP; see Figure 2.12 [61]. The former garden chairs can now even be used for the same application. The embrittlement can be reduced further when additional antioxidants are added. This shows the huge potential of stabilizers, but also the complexity of selection and combination of these additives.

The mechanical recycling of postconsumer waste leads inevitably to the formation of polymer blends and most common polymers are immiscible in the melt phase. This usually results in lower mechanical performance, often referred to as downcycling. The reason is phase separation and low adhesion between the two or more phases, which leads most commonly to low impact strength. The solution for virgin blends can be applied to blends of recycled plastics as well: compatibilizers.

Compatibilizers for blends from recycled plastics are available for polyolefin blends of PP and PEs, polyolefins contaminated with polar polymer (e.g., PA) from plastic films, and PET or PA contaminated with low amounts of PEs or PP (from bottle caps). Figure 2.13 shows the effect of the compatibilizer SEBS-g-MAH on the otherwise immiscible blend of PP/PET. Both the elongation at break and the impact strength increase with the additive content, highlighting the ductile behavior of the now miscible blend [61].

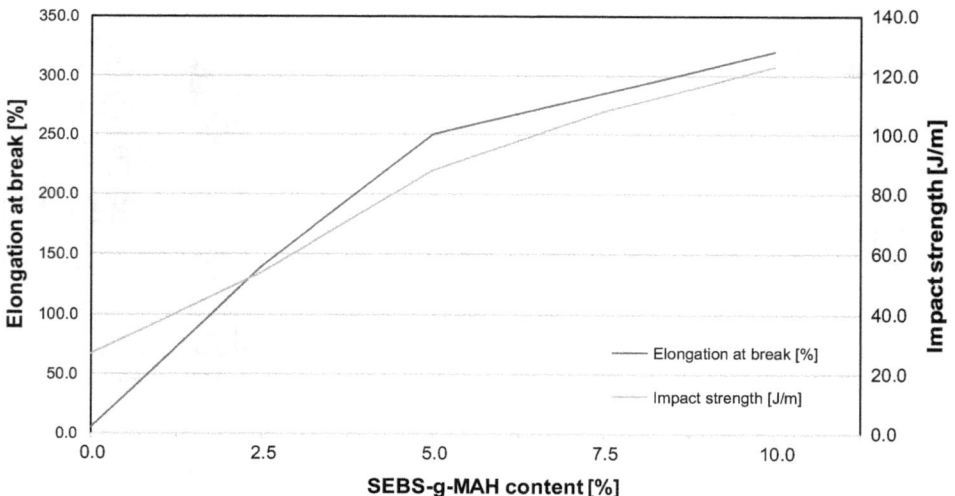

Figure 2.13 Effect of compatibilizers as illustrated by a PP/PET blend [61]

2.4.4.2 Fiber-Reinforced Plastics

Dr. Sebastian Goris

Fiber-reinforced plastics (FRP) are a distinctive class of materials defined by the combination of at least two constituents: the matrix material and the reinforcing fibers. By merging the separate phases, the composite has enhanced properties that exceed the performance and capabilities of the individual constituents. Fiber-reinforced composites are classified into *continuous* and *discontinuous* fiber-reinforced composites. Fiber-reinforced composites are commonly applied as a substitute for metal as they provide key advantages in form of lower manufacturing costs, component integration, lower density, and high-volume production.

Continuous fibers for reinforcement can be used in many forms, such as woven, fabrics, knitted constructions, or laminated composites. Laminated composites consist of multiple layers of unidirectional fibers each oriented differently, to optimize performance at laminate level. Laminated composite materials are commonly used for high-performance applications, because of their superior mechanical properties.

Discontinuous fibers are another form factor of the fibers within the composite in which fibers are used to improve the performance of the base resin. This material class is characterized by the fact that the fibers are chopped and experience a change in orientation and a reduction in length during molding. This process-induced change in the fiber microstructure determines the performance of the molded part.

FRP materials are used in various industries, such as automotive, aerospace, construction, consumer goods, sporting goods, oil and gas, and wind power. Their utilization has seen continuous growth, particularly for applications that replace metals for lightweight design. The recent increase in production due to the many applications will inevitably result in an increase in the amount of end-of-life material and waste. Hence, the question of recycling for FRP has become a major topic of research in academia and industry. While most challenges in the recycling of unfilled plastics also apply to FRP, there are additional considerations for recycling fiber-reinforced materials. In particular, the reduction of the length of the fibers through additional reprocessing can have a profound impact on the value of recycled FRP parts.

2.4.4.2.1 Discontinuous Fiber-Reinforced Composites

Discontinuous fiber-reinforced composites comprise either a *thermoset* or a *thermoplastic* matrix. Sheet molding compound (SMC) is the most prominent type of discontinuous fiber-reinforced thermoset and is processed in compression molding [49]. A variety of thermoplastic matrices and processes are available for discontinuous fiber-reinforced composites, while engineering thermoplastics, such as polyamide (PA) or polypropylene (PP), represent the majority of matrices [50]. Glass fibers are frequently used for reinforcement due to their availability, low cost, and high strength. Although carbon fibers can offer superior performance compared to glass fibers, the cost increase often does not justify their use as substitute in FRP for all applications [51].

Nevertheless, recycling of carbon fibers has become an active field of research aiming to use scrap fiber material from continuous carbon fiber processes and reuse it in discontinuous fiber-reinforced compounds [52].

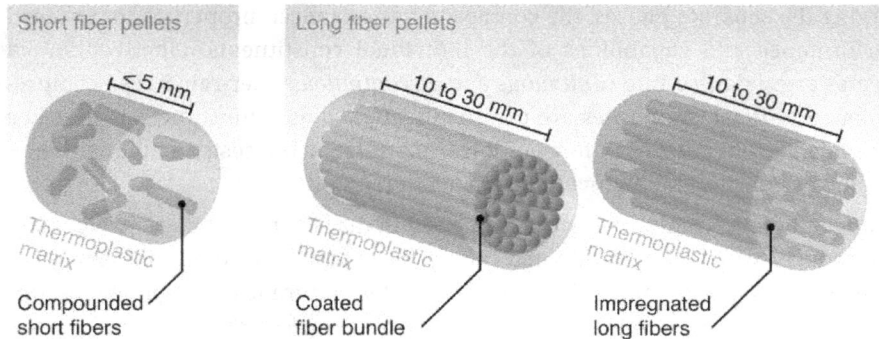

Figure 2.14 Illustration of short fiber pellets (left) and long fiber pellets, which are further divided in coated fiber bundles (center) and pultruded fibers (right) [49]

Discontinuous fiber-reinforced thermoplastics can be further classified as *short fiber-reinforced thermoplastics* (SFT) and *long fiber-reinforced thermoplastics* (LFT). Both classes are delivered in pellet form, as schematically depicted in Figure 2.14. However, the two types are differentiated in the initial length of the fibers within the pellet before processing. Another way to look at the distinction is the average fiber length to fiber diameter ratio, also called the *fiber aspect ratio*. If the ratio is below 100, the composite is commonly classified as an SFT. If the value exceeds 100, it is referred to as an LFT. The performance and cost of LFT compounds usually place them between continuous fiber-reinforced composites used for high-performance applications and SFC compounds.

2.4.4.2.2 Fiber Length Reduction

The reinforcing characteristic of the fibers is achieved by transferring the applied load from the matrix to the fibers. Consequently, the interfacial adhesion as well as the length of the fibers embedded in the matrix determine the maximum transferable load and the overall reinforcement. The effect of the residual length of the fibers on the mechanical properties of a molded part is illustrated in Figure 2.15. The graph shows the effect of the fiber aspect ratio on the tensile modulus, tensile strength, and impact strength [53, 54, 55, 56]. The figure also qualitatively illustrates the expected range of the residual fiber aspect ratios in LFT injection molding, as reported in the literature [49].

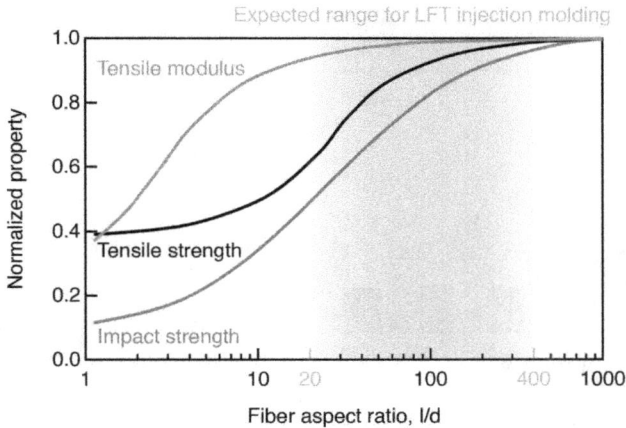

Figure 2.15
Normalized mechanical properties as a function of fiber aspect ratio for tensile modulus, tensile strength, and impact strength. The graph is adapted from Thomason et al. [53, 54, 55, 56]. The highlighted range illustrates the fiber aspect expected in LFT injection molding

Preserving a fiber length that is sufficiently long to transfer the maximum load, while maintaining economical process cycles, is a major challenge for molding SFT and LFT materials. While the initial length of the fibers may be uniform and up to 30 mm for an LFT compound, fiber attrition results in a distribution of fiber lengths in the finished part after only one processing event ranging from powder-like fibers (< 0.1 mm) to a small fraction of fibers retaining their initial length. In general, the degree of fiber attrition is influenced by the process, processing parameters, material selection, and mold design [63]. On the micro-scale, the driving mechanisms for fiber attrition are hydrodynamic effects, fiber–fiber interactions, and fiber–equipment interactions, which the fibers experience along all stages of the process. For a comprehensive review of the underlying physics of fiber attrition during processing, the reader is referred to [57] and [49].

Lafranche et al. [58] conducted a comprehensive experimental study to evaluate the fiber length reduction along all stages of a single injection molding cycle. Figure 2.16 shows the measured fiber length from the pellets (L_N = 7.9 mm) to the molded part (L_N = 1.9 mm), which is an overall fiber length reduction of more than 75%. It can be seen that the majority of the length reduction happens in the screw sections and only small changes occur in the runner and cavity. Furthermore, this shows that the improvement in mechanical performance of injection molded FRP can be mainly attributed to a fiber length much shorter than in the pellet.

Further studies indicate that the residual fiber length varies substantially within a molded part [49, 57]. Depending on the material properties, fiber concentration, processing conditions, and mold geometry, the local fiber length within the molded part can vary up to 30% [59]. This finding further highlights the heterogeneous fiber microstructure seen in discontinuous fiber-reinforced thermoplastics and it shows that the common assumption of a uniform fiber length throughout a molded part is not valid. Nevertheless, the majority of the fiber length reduction occurs during the plastification stage.

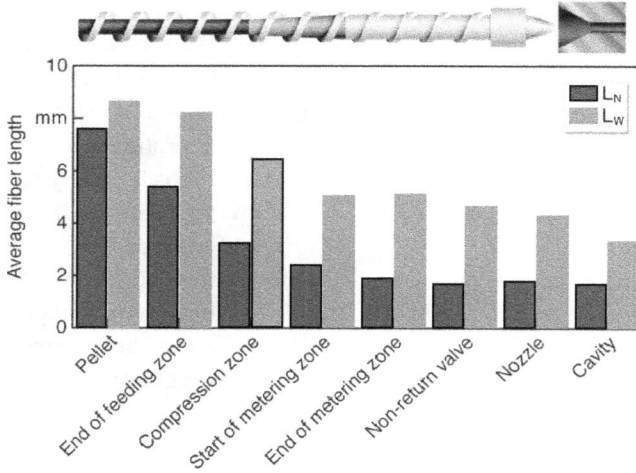

Figure 2.16 Fiber breakage during the plasticizing stage, adapted from [58]. L_N is the number-average fiber length and L_W is the weight-average fiber length. While L_N represents the mean of the measured fiber population, the characteristic of L_W is a weighted average emphasizing the proportion of long fibers in the distribution

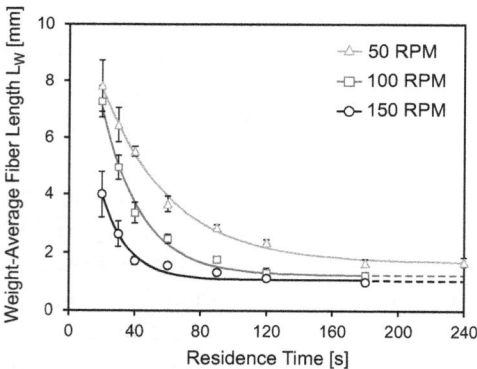

Figure 2.17
Fiber length reduction for glass fiber-reinforced polypropylene under steady shear, adapted from [57]. A 30 wt% filled polypropylene was processed under iso-thermal melt temperature in a custom-ized Couette rheometer. The residual fiber length was measured at varying residence times to understand the evo-lution of fiber attrition

A study of fiber length reduction under controlled conditions using steady shear in a customized Couette rheometer showed the continuous reduction of fiber length in processing LFT compounds [57]. Polypropylene at 30 wt% fiber loading was exposed to steady shear conditions under isothermal melt temperature (250 °C). As shown in Figure 2.17, the fiber length rapidly decreases from 15 mm down to approximately 1.6 mm for a speed of 50 RPM or 1 mm for a speed of 150 RPM. Even at a low residence time (< 60 seconds), the residual fiber length is only approximately 27% of the initial fiber length. This highlights the sensitivity of fiber attrition and processing, which consequently poses a major challenge in recycling fiber-reinforced composites. From this experimental study, it can be inferred that an *unbreakable fiber length* exists, at which point no additional fiber breakage occurs for a given processing condition [57]. While the correlation between the idealized conditions in the Couette rheometer and conventional molding processes is still subject to research efforts, [57] the work suggests that there is a lower bound on residual fiber length that is reached in recycling of discontinuous fiber-reinforced plastics.

2.4.4.2.3 Recycling of Fiber-Reinforced Composites

The premium paid for the reinforcing fibers over unfilled resins, particularly for more expensive fibers such as carbon fibers, warrants the consideration of recycling the fiber-reinforced composites. While the application of this material class is increasing, recycling and end-of-life considerations remain a mostly academic research topic. Major challenges in recycling remain, mostly attributed to the fact that fiber-reinforced composites are made from a composition of several constituents with differing material characteristics. For the most simplistic fiber-reinforced composite, the constituents include at least the fibers, the matrix, and the fiber sizing. For now, complete separation and recycling of the individual components is challenging, if not impossible, in terms of economic considerations.

Additionally, fiber length reduction is a major challenge since the average fiber length is a key determinant of the mechanical performance of the molded part. Any reprocessing of fiber-filled compounds can result in additional reduction of fiber length and, consequently, lower the mechanical performance. Studies [39, 40] show that the recycling of manufacturing scrap leads to significant deterioration in the mechanical properties compared to the virgin material, which has been attributed to the additional reduction in fiber length, as shown in Figure 2.18. While the change in properties is substantial, it can be seen that refreshing of recycled material with virgin material can offset the loss in performance significantly, as shown in Figure 2.19.

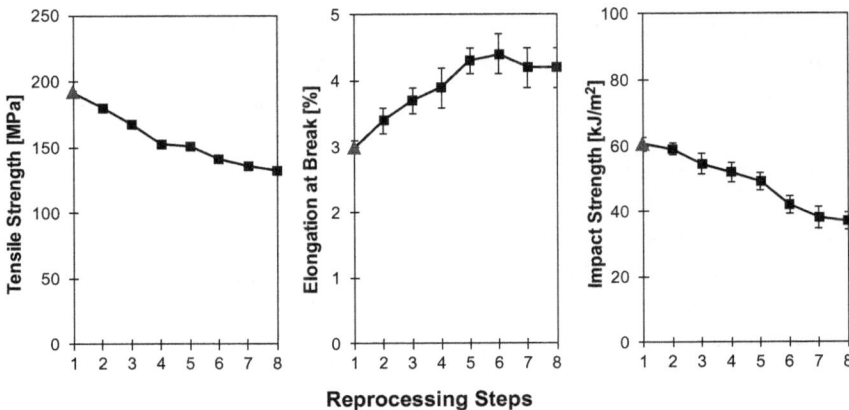

Figure 2.18 Mechanical properties of recycled fiberglass/PA66 composite as a function of the number of reprocessing steps (injection molding) [47, 48]

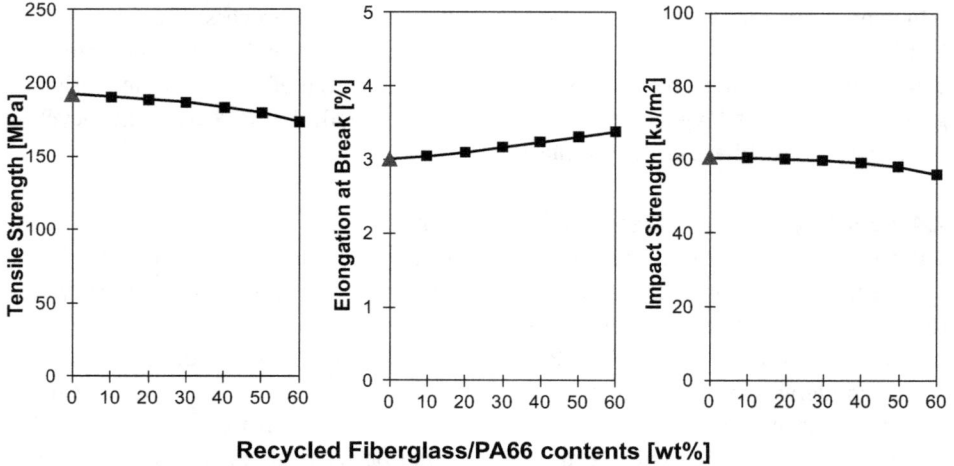

Recycled Fiberglass/PA66 contents [wt%]

Figure 2.19 Mechanical properties of fiberglass/PA66 composite as a function of the percentage of recycled composite, in the first reprocessing step [47, 48]

As described in the previous section, the majority of the fiber attrition occurs inside the screw during plastification and it can be assumed that a lower bound of fiber length reduction (unbreakable fiber length) exists for given processing conditions. Hence, comparing the recycling of preconsumer to postconsumer waste, no significant difference in the final fiber lengths are expected as most of the fiber length reduction occurs during plastification.

2.5 Contaminants

Plastics are hard to clean due to the penetration of contaminants into the polymer matrix. Composites and mixed plastic waste are especially difficult to separate into the different plastic types, which all require different reprocessing techniques and settings [31].

There are a number of contaminants that can significantly obstruct the recycling process and result in severe deterioration of performance of recycled material. During the extrusion process in which the recycled material is subjected to high temperatures and mechanical stresses, the presence of contaminants may lead to hydrolytic and thermal degradation and subsequent decreases in both the molecular weight and viscosity of the plastic.

In the case of the PET recycling process, the common contaminants are acids and acid-producing compounds, for example, which arise when PET and PVC are mixed. Hydrochloric acid produced from PVC acts as catalyst for chain-cleavage reactions. Likewise, elevated water content can lead to chain cleavage through hydrolysis. Most water contamination comes from the washing process and can be removed by proper drying. Dyes and coloring agents are another source of contamination, leading to undesirable, mostly brownish color in the recycled plastic. Contaminants, such as acetaldehyde (a natural degradation product of PET) and other contaminants arising from misuse of PET by consumers (such as for storage of fuel, pesticides, and other dangerous materials), are potential health hazards in recycled PET products [31].

Finally, odors are a major problem for recyclate adoption, which primarily stems from contaminants. Odors in recycled plastics often originate from residual substances, microbial activity, chemical additives, and cross-contamination. Packaging materials can retain traces of their original contents, such as food, cleaning agents, or chemicals, which lead to persistent odors in recyclates. Additionally, organic residues left on plastics can undergo microbial decomposition, particularly in the presence of moisture, generating malodorous compounds. The recycling process itself contributes to odor formation through the release of volatile organic compounds (VOCs) from plastic additives like stabilizers, plasticizers, and flame retardants, as well as degradation by-products. Cross-contamination further exacerbates the issue, as the unintended mixing of polymers, such as PVC in PET streams, can result in the release of hydrochloric acid and other reactive compounds, degrading the material and intensifying unwanted smells.

2.6 Conclusion: Technical Feasibility of Plastics Recycling

In summary, the recyclability of plastic waste depends on the origin of the waste as well as the sensitivity of the polymer(s) to degradation. In most cases, preconsumer waste (manufacturing scrap) can be reprocessed with little deterioration of properties. The property changes can even be minimized or extended by refreshing the regrind waste with virgin plastic. Although the processing properties such as viscosity are affected, the changes in service-life properties such as mechanical performance are often negligible. In industrial production, the regrind proportion can be as high as 70%.

The recycling rate of postconsumer waste is increasing at a slow pace. This is due to technical limitations such as the limited availability of clean and unmixed postconsumer plastic waste. This rate can only be increased with improved sorting processes as well as when the recycling process becomes an integral part of the product design process and both the manufacturer and consumer take an active part in the improvement of that process. While mechanical recycling is possible with PCR, the most promising approaches for postconsumer waste to date are in chemical recycling. This is especially true for the fractions that are too contaminated for mechanical recycling.

References

[1] The Association of Postconsumer Plastic Recyclers and American Chemistry Council. 2013 *United States National Post-consumer Plastics Bottle Recycling Report*. Washington, D. C. pp. 1–10, 2014.

[2] American Chemistry Council Economics & Statistics Department. *Plastic Resins in the United States*. Washington, D. C. 2013.

[3] Hopmann, C., and Michaeli, W. *Einführung in die Kunststoffverarbeitung*, 7th ed. Hanser, Munich. 2015.

[4] Ignatyev, I. A., Thielemans, W., and Vander Beke, B. Recycling of polymers: a review. *ChemSusChem*. vol. 7, pp. 1579–1593, 2014.

[5] Hamad, K., Kaseem, M., and Deri, F. Recycling of waste from polymer materials: an overview of the recent works. *Polymer Degradation and Stability*. vol. 98, pp. 2801–2812, 2013.

[6] Triebert, D., Hanel, H., Bundt, M., Wohnig, K. Solvent-Based Recycling. In Circular Economy of Polymers: Topics in Recycling Technologies, *ACS Symposium Series*, pp. 33–59, 2021.

[7] Rahimi, A., García, J. M. Chemical recycling of waste plastics for new materials production. Nature Reviews Chemistry, vol. 1 (6), 2017.

[8] Geyer, R., Jambeck, J. R., Law, K. L. Production, use, and fate of all plastics ever made. *Science Advances*, vol. 3, 2017.

[9] Muszynski, M. Nowicki, J. Zygadlo, M. Dudek, G. Comparsion of Catalyst Effectiveness in Different Chemical Depolymerization Methods of Poly(ethylene terephthalate). *Molecules*, vol. 28 (17), 2023.

[10] Li, H., Aguirre-Villegas, H. A., Allen, R. D., Bai, X., Benson, C. H., Beckham, G. T., Bradshaw, S. L., Brown, J. L., Brown, R. C., Cecon, V. S., et al. Expanding plastics recycling technologies: chemical aspects, technology status and challenges. *Green Chemistry*, vol. 24 (23), pp. 8899–9002, 2022.

[11] Enache, A. C., Grecu, I., Samoila, P. Polyethylene Terephthalate (PET) Recycled by Catalytic Glycolysis: A Bridge toward Circular Economy Principles. *Materials (Basel)*, vol. 17 (12), 2024.

[12] Sun, H., Chen, Z., Zhou, J., Chen, L., Zuo, W. Recovery of high-quality terephthalic acid from waste polyester textiles via a neutral hydrolysis method. *Journal of Environmental Chemical Engineering*, vol. 12 (3), 2024.

[13] Cao, J., Liang, H., Yang, J., Zhu, Z., Deng, J., Li, X., Elimelech, M., Lu, X. Depolymerization mechanisms and closed-loop assessment in polyester waste recycling. *Nat. Commun.*, vol. 15 (1), p. 6266, 2024.

[14] Muringayil Joseph, T., Azat, S., Ahmadi, Z., Moini Jazani, O., Esmaeili, A., Kianfar, E., Haponiuk, J., Thomas, S. Polyethylene terephthalate (PET) recycling: A review. *Case Studies in Chemical and Environmental Engineering*, vol. 9, 2024.

[15] Wu, Y., Hu, Q., Che, Y., Niu, Z. Opportunities and challenges for plastic depolymerization by biomimetic catalysis. *Chem. Sci.*, vol. 15 (17), pp. 6200–6217, 2024.

[16] Afzal, S., Singh, A., Nicholson, S. R., Uekert, T., DesVeaux, J. S., Tan, E. C. D., Dutta, A., Carpenter, A. C., Baldwin, R. M., Beckham, G. T. Techno-economic analysis and life cycle assessment of mixed plastic waste gasification for production of methanol and hydrogen. *Green Chemistry*, vol. 25 (13), pp. 5068–5085. 2023.

[17] Goodship, V. *Introduction to Plastics Recycling*. Smithers Rapra, Shawbury, United Kingdom. 2007.

[18] Brandrup, J., Bittner, M., Michaeli, W., and Menges, G. *Die Wiederverwertung von Kunststoffen*. Hanser, Munich. pp. 738–745, 1995.

[19] United States Environmental Protection Agency (EPA). *Advancing Sustainable Materials Management: Facts and Figures 2017. Assessing Trends in Material Generation, Recycling, Composting, Combustion with Energy Recovery and Landfilling in the United States*. Washington, D. C. 2019.

[20] Vault Consulting, LLC. *Resin Sales and Captive Use by Major Market*. Prepared for American Chemistry Council Plastics Industry Producers Statistics' Group. 2016.

[21] European Commission's Directorate-General Environment. *Plastic Waste: Ecological and Human Health Impacts*. Science for Environment Policy, In-depth Report. pp. 1–37, 2011.

[22] Franklin Associates. *Impact of Plastics Packaging on Life Cycle Energy Consumption & Greenhouse Gas Emissions in the United States and Canada*. Prepared for American Chemistry Council (ACC) and Canadian Plastics Industry Association (CPIA). 2014.

[23] Gerard, K. *Use of Plastic Materials in the Construction Industry*. Access Date: 2016/11/08. Available: *http://info.craftechind.com/blog/bid/380434/Use-of-Plastic-Materials-in-the-Construction-Industry*.

[24] *nibusinessinfo.co.uk. Recycling construction materials – Recycling plastic from construction projects*. Access Date: 2016/11/08. Available: *https://www.nibusinessinfo.co.uk/content/recycling-plastic-con struction-projects*.

[25] Thibodeaux, J. *Recycling agricultural plastic is growing industry*. Access Date: 2016/11/08. Available: *http://www.greensourcedfw.org/articles/recycling-agricultural-plastic-growing-industry*.

[26] Grossmann, E. *How can agriculture solve its $5.87 billion plastic problem?* Access Date: 2016/11/08. Available: *https://www.greenbiz.com/article/how-can-agriculture-solve-its-1-billion-plastic-problem*.

[27] Barth, B. *3 Ways Farmers are Kicking the Plastic Habit*. Access Date: 2016/11/08. Available: *http:// modernfarmer.com/2015/09/agriculture-plastic-waste/*.

[28] Schwesig, A., Riise, B., and Ybbs, D. *PC/ABS recovered from shredded waste electrical and electronics equipment*. Paper presented at SPE Annual Technical Conference (ANTEC), Indianapolis, IN, 2016. In SPE ANTEC Conference Proceedings. pp. 1774–1779.

[29] Plastic Moulding. *Polyethylene*. Access Date: 2016/11/08. Available: *http://www.plasticmoulding.ca/ polymers/polyethylene.htm*.

[30] Polymer Properties Database. *A-Z Polymer Data*. Access Date: 2016/11/08. Available: *http://polymer database.com/home.html*.

[31] Holton, C. Dissolving the plastics problem. *Environmental Health Perspectives*. vol. 105, pp. 388–390, 1997.

[32] Illumin. *Recycling Plastics: New Recycling Technology and Biodegradable Polymer Development*. Access Date: 2016/11/08. Available: *http://illumin.usc.edu/7/recycling-plastics-new-recycling-technol ogy-and-biodegradable-polymer-development/*.

[33] Ruj, B., Pandey, V., Jash, P., and Srivastava, V. Sorting of plastic waste for effective recycling. *International Journal of Applied Science and Engineering Research*. vol. 4, pp. 564–571, 2015.

[34] Sanchez-Parra, K. M. *Effects of processing events on commodity thermoplastics*. Thesis, University of Wisconsin–Madison, Madison, WI. 2013.

[35] Basedow, A. A. M., Ebert, K. H., and Hunger, H. Effects of mechanical stress on the reactivity of polymers: shear degradation of polyacrylamide and dextran. *Die Makromolekulare Chemie*. vol. 180, pp. 411–427, 1979.

[36] Goldberg, V. M., and Zaikov, G. E. Kinetics of mechanical degradation in melts under model conditions and during processing of polymers—a review. *Polymer Degradation and Stability*. vol. 19, pp. 221–250, 1987.

[37] Peacock, A. J., and Calhoun, A. *Polymer Chemistry: Properties and Applications*. Hanser Gardner, Cincinnati, OH. 2006.

[38] Schmiederer, D. *Schonende Spritzgiessverarbeitung von Thermoplasten*: Univ. Erlangen-Nürnberg, Lehrstuhl für Kunststofftechnik. 2008.

[39] Zweifel, H. *Stabilization of Polymeric Materials*. Springer-Verlag, Berlin, Germany. 1998.

[40] Gugumus, F. Mechanisms of thermooxidative stabilization with HAS. *Polymer Degradation and Stability*. vol. 44, pp. 299–322, 1994.

[41] Moss, S., and Zweifel, H. Degradation and stabilization of high density polyethylene during multiple extrusions. *Polymer Degradation and Stability*. vol. 25, pp. 217–245, 1989.

[42] Frounchi, M. Studies on degradation of PET in mechanical recycling. *Macromolecular Symposia*. vol. 144, pp. 465–469, 1999.

[43] DeSousa, J. The effects of multiple heat histories on the mechanical properties of high-impact polystyrene. Paper presented at SPE Annual Technical (ANTEC), San Francisco, CA, May, 2002. In *SPE ANTEC Conference Proceedings*. vol. 3, p. 2925.

[44] Vilaplana, F., Ribes-Greus, A., and Karlsson, S. Degradation of recycled high-impact polystyrene. Simulation by reprocessing and thermo-oxidation. *Polymer Degradation and Stability*. vol. 91, pp. 2163–2170, 2006.

[45] Yarahmadi, N., Jakubowicz, I., and Enebro, J. Polylactic acid and its blends with petroleum-based resins: effects of reprocessing and recycling on properties. *Journal of Applied Polymer Science.* vol. 133, pp. 1–9, 2016.

[46] Case, R. M., Korzen, A. P., and Maclean, S. B. The effects of regrind loading levels and heat history on the properties of selected engineering polymers. Paper presented at SPE Annual Technical Conference (ANTEC), Dallas, TX, May 6–16. 2001.

[47] Eriksson, P.-A., Albertsson, A.-C., Boydell, P., Eriksson, K., and Månson, J.-A. E. Reprocessing of fiberglass reinforced polyamide 66: influence on short term properties. *Polymer Composites.* vol. 17, pp. 823–829, 1996.

[48] Eriksson, P. A., Albertsson, A. C., Boydell, P., Prautzsch, G., and Manson, J. A. E. Prediction of mechanical properties of recycled fiberglass reinforced polyamide 66. *Polymer Composites.* vol. 17, pp. 830–839, 1996.

[49] Gandhi, U., Goris, S., Osswald, T. A., and Song, S. *Discontinuous Fiber-Reinforced Composites: Fundamentals and Applications.* Hanser, Munich. 2020.

[50] I. V. K. AVK, *Handbuch Faserverbundkunststoffe – Grundlagen, Verarbeitung, Anwendungen.* Vieweg+ Teubner Verlag, Wiesbaden, Germany. 2010.

[51] Vaidya, V. *Composites for Automotive, Truck and Mass Transit – Materials, Design, Manufacturing.* DEStech Publications, Inc., Pennsylvania, U.S. 2011.

[52] Albrecht, K. *Nachhaltige faserverstaerkte Kunststoffe im Spritzguss – Faserorientierung und Faserschaedigung im Experiment und Simulation,* Ph.D. thesis, University of Erlangen-Nuremberg, 2018.

[53] Thomason, J. L., and Vlug, M. A. Influence of fibre length and concentration on the properties of glass fibre-reinforced polypropylene: 1. Tensile and flexural modulus. *Compos. Part Appl. Sci. Manuf.* vol. 27, no. 6, pp. 477–484, 1996.

[54] Thomason, J. L. The influence of fibre length, diameter and concentration on the impact performance of long glass-fibre reinforced polyamide 6,6. *Compos. Part Appl. Sci. Manuf.* vol. 40, no. 2, pp. 114–124, 2009.

[55] Thomason, J. L., Vlug, M. A., Schipper, G., and Krikor, H. G. L. T. Influence of fibre length and concentration on the properties of glass fibre-reinforced polypropylene: Part 3. Strength and strain at failure. *Compos. Part Appl. Sci. Manuf.* vol. 27, no. 11, pp. 1075–1084, 1996.

[56] Thomason, J. L., and Vlug, M. A. Influence of fibre length and concentration on the properties of glass fibre-reinforced polypropylene: 4. Impact properties. *Compos. Part Appl. Sci. Manuf.* vol. 28, no. 3, pp. 277–288, 1997.

[57] Goris, S. *Characterization of the Process-Induced Fiber Configuration of Long Glass Fiber-Reinforced Thermoplastics,* Ph.D. thesis, University of Wisconsin-Madison.

[58] Lafranche, E., Krawczak, P., Ciolczyk, J.-P., and Maugey, J. Injection moulding of long glass fiber reinforced polyamide 66: Processing conditions/microstructure/flexural properties relationship. *Adv. Polym. Technol.* vol. 24, no. 2, pp. 114–131, 2005.

[59] Goris, S., Back, T., Yanev, A., Brands, D., Drummer, D., and Osswald, T. A. A novel fiber length measurement technique for discontinuous fiber-reinforced composites: A comparative study with existing methods. *Polymer Composites.* 39, pp. 4058–4070, 2018. doi:10.1002/pc.24466.

[60] Achilias, D. S., and Antonakou, E. V. Chemical and thermochemical recycling of polymers from waste electrical and electronic equipment, in *Recycling Materials Based on Environmentally Friendly Techniques,* IntechOpen, 2015.

[61] Pfaendner, R. Improving the quality of recycled materials – an overview of suitable additives, *Kunststoffe International,* 12/2015. pp. 41–44.

[62] Pospisil, J., Sitek, F. A., and Pfaendner, R. Upgrading of recycled plastics by restabilization – an overview. *Polymer Degradation and Stability,* 48, pp. 351–358, 1995.

[63] Osswald, T. A., and Menges, G. *Materials Science of Polymers for Engineers,* 3rd ed. Hanser, Munich. 2012.

3

Quality of Recyclates

3.1 Testing Methods for Recyclates

3.1.1 Introduction: Why Testing Is Critical for Recyclates

Polymer recyclates contain a mix of degraded polymers, fillers, additives, and contaminants, requiring comprehensive testing to ensure consistent quality and performance. Unlike virgin materials, recyclates show higher variability, making frequent testing essential for processability, mechanical strength, and regulatory compliance. These tests help determine recyclates' fitness for specific processes such as injection molding, extrusion, and blending with virgin materials or other recyclate batches, while also assessing long-term durability and safety.

One challenge in recycling is the inaccurate sorting of plastic waste, which can lead to unintended polymer mixtures in recyclates. These unintentional blends can significantly affect mechanical, thermal, and rheological properties, making extensive testing necessary to evaluate their behavior during processing and final application performance.

3.1.2 Categories of Testing Methods

To fully characterize recyclates, testing methods are categorized into:

- **Thermal analysis** – Evaluates melting behavior, crystallinity, degradation, and identifies (polymer) composition and contaminants.

- **Spectroscopy & chemical analysis** – Identifies polymer composition and contaminants.

- **Rheology** – Determines flow behavior and processability.

- **Mechanical testing** – Measures strength, toughness, and durability.

- **Optical & structural analysis** – Examines surface morphology and defects.

- **Environmental stress cracking (ESR)** – Assesses material performance under stress and chemical exposure.

3.1.3 Thermal Analysis

Differential scanning calorimetry (DSC) measures heat flow during glass transition, melting and crystallization, providing key insights into polymer composition, purity, and crystallinity. It is particularly useful for distinguishing blends in mixed recyclates, such as LDPE, HDPE, and PP. DSC also helps assess thermal history and processing conditions.

Oxidative induction time (OIT) and Oxidative onset temperature (OOT) are critical for evaluating the oxidative stability of recyclates. OIT measures the time a polymer can withstand an oxygen-rich environment at an elevated temperature before significant oxidation occurs, which is important for predicting long-term durability. OOT, on the other hand, determines the temperature at which oxidation begins. Both metrics are critical for assessing the remaining stabilizer content of recycled polymers.

Thermogravimetric analysis (TGA) quantifies polymer and filler content by measuring weight loss as a sample is heated. This helps identify contaminants like cellulose, ink residues, and inorganic fillers. Coupling TGA with **evolved gas analysis (EGA)** using FTIR or GC-MS allows deeper analysis of decomposition products, such as volatile additives or contaminants.

Temperature rising elution fractionation (TREF) and **crystallization analysis fractionation (CRYSTAF)** are specialized techniques for analyzing polyolefin recyclates, helping to differentiate fractions based on crystallinity and molecular weight distribution. These methods are particularly useful for assessing variations within mixed polyolefin streams.

3.1.4 Spectroscopy and Chromatography

Fourier-transform infrared spectroscopy (FTIR) identifies polymers by analyzing their molecular vibrations upon infrared light absorption, producing a unique spectral fingerprint for each material. It is widely used to differentiate common polymers such as PE, PP, PET, and PVC, and to detect certain additives or contaminants. However, FTIR has limitations in recycling applications, particularly with carbon-black-filled or highly pigmented plastics, which absorb infrared light and reduce spectral clarity. It also struggles to distinguish chemically similar polymers like LDPE and HDPE or to differentiate copolymers from blends, making it less effective for detailed recyclate characterization in complex waste streams.

Near-infrared spectroscopy (NIR) operates in a different part of the spectrum than FTIR and is widely used for high-speed polymer sorting in recycling facilities. Unlike FTIR, NIR can be effectively applied in automated sorting lines, where it quickly differentiates between major polymer types. However, like FTIR, it also struggles with dark or carbon-black-containing materials and is less precise for distinguishing chemically similar polymers.

High-performance liquid chromatography (HPLC) is employed to detect additives, degradation products, and residual monomers within recyclates. This is particularly important for food-contact applications and regulatory compliance, where even trace amounts of contaminants must be identified and quantified. HPLC enables the detection of stabilizers, plasticizers, and other small molecules that impact the performance and safety of the final recycled product.

Size exclusion chromatography (SEC), also known as **gel permeation chromatography (GPC)**, is used to determine the molecular weight distribution of polymers, a crucial factor in assessing polymer degradation after multiple processing cycles. By analyzing the chain length and molecular weight of recyclates, SEC helps predict whether a material will retain its mechanical and processing properties after reprocessing.

3.1.5 Rheology

Understanding the flow behavior of recyclates is critical for ensuring they can be processed efficiently in extrusion, injection molding, and film production. Rheological testing of recyclates evaluates their shear viscosity, elongational viscosity, melt flow properties, and viscoelastic behavior – all of which can vary significantly depending on the presence of contaminants, additives, and degraded polymer chains.

MFI is a widely used, quick method for estimating polymer flow behavior under a specific load and temperature. It is commonly applied in quality control but has limitations – it provides only a single-point measurement and does not account for how viscosity changes under different shear rates. **VN**, measured via dilute solution viscometry, gives insight into the molecular weight of the polymer, which influences mechanical properties and processability. VN is particularly relevant for recyclates used in applications requiring controlled molecular weight

Capillary rheometers measure viscosity under high shear rates, replicating conditions found in industrial polymer processing. This is essential for assessing how a recyclate behaves during extrusion or molding, identifying shear thinning, and determining whether it can replace virgin material. Capillary rheometry also provides extensional viscosity data, relevant for applications like film blowing and fiber spinning.

Rotational and **oscillatory rheometry** are performed using the same instrument, but under different testing conditions. Rotational rheometry is used to characterize recyclates under low shear conditions, offering insights into melt elasticity, stability, and structural recovery. This method is particularly useful for determining batch-to-batch variations in recyclates, which can affect consistency in manufacturing processes.

Oscillatory rheometry, on the other hand, provides detailed information about viscoelastic properties, which are critical for understanding polymer weight changes, compatibilization of blends (mixed polymer types found in recyclates), and additive performance. This technique involves subjecting the polymer melt to small, oscillatory deformations, measuring the material's storage modulus (G') and loss modulus (G'').

- **Han plots** help assess the miscibility of polymer blends by plotting G' against G''. If the plot follows a straight line, the blend is well-mixed, whereas deviations indicate phase separation or incompatibility, which can lead to poor recyclate performance.

- **Tan δ** (damping factor), which is the ratio of loss modulus to storage modulus (G''/G'), provides insight into the molecular relaxation behavior of a material. A high Tan δ indicates a more viscous or damping material, while a low value suggests a more elastic response. This is particularly useful for understanding whether recycled polymers have retained their flexibility or become brittle due to thermal degradation or contamination.

Together, these rheological techniques provide a complete picture of recyclate behavior, ensuring it meets processing and product performance requirements.

3.1.6 Mechanical Testing

Mechanical testing is essential for determining the strength, durability, and suitability of recycled polymers for various applications. Recyclates can experience degradation, leading to weakened mechanical properties compared to virgin materials. Several standardized mechanical tests help assess how well a recycled polymer can perform in its intended use.

Tensile testing measures a material's tensile strength, elongation at break, and Young's modulus, providing a direct assessment of how much stress a polymer can endure before failure. It is a crucial test for recyclates, as polymer degradation or contamination can significantly impact mechanical properties. Differences in tensile properties between batches of recycled materials can indicate inconsistencies in processing, unintentional polymer blending, or contamination levels.

Impact tests evaluate the ability of a material to withstand sudden forces without breaking. Charpy and Izod impact tests measure how much energy a material absorbs before fracturing, which is particularly important for automotive, packaging, and structural applications. Recyclates tend to have lower impact resistance due to molecular chain scission and contamination, making this test valuable for quality control.

Creep and Stress Relaxation tests assess a polymer's ability to withstand constant mechanical stress over time. Creep testing measures how much a material deforms under sustained load, whereas stress relaxation testing observes how internal stresses reduce over time. For recycled materials, these tests are critical in applications where long-term structural integrity is necessary, such as construction or transportation components.

Dynamic Mechanical Analysis (DMA) applies cyclic loads to a solid polymer sample to determine its viscoelastic behavior, making it a powerful tool for accelerated creep testing. Unlike conventional creep testing, which can take weeks or months to show meaningful results, DMA enables long-term predictions within a few days by extrapolating data from short-term dynamic stress tests. This capability is particularly useful for assessing how recyclates will perform under prolonged mechanical loads, such as in construction materials, automotive parts, and structural applications.

By measuring storage modulus (G') and loss modulus (G''), DMA helps predict whether a polymer will maintain its mechanical integrity over time or experience excessive deformation. It also provides insight into temperature-dependent mechanical performance, allowing for a better understanding the heat deflection limits of recyclates.

3.1.7 Optical Techniques

Understanding the microstructure of recyclates helps identify defects, phase separation, and filler distribution, which can impact mechanical and processing properties. Microscopy techniques provide detailed visual insights into a polymer's surface and internal composition.

Scanning Electron Microscopy (SEM) provides high-resolution images of a polymer's surface structure, revealing defects like cracks, voids, or phase separation. This is particularly useful for assessing recyclates, which may contain traces of previous use, contamination, or degradation-induced defects.

Light microscopy, although less powerful than SEM, is a quick and effective method for initial visual inspections, such as detecting large inclusions, poor dispersion of fillers, or uneven morphology in a recycled polymer sample.

3.1.8 Environmental Stress Cracking (ESR)

Environmental Stress Cracking (ESR) is a common failure mode in polymers where a combination of mechanical stress and chemical exposure leads to microcrack formation and eventual material failure. This is especially relevant for recyclates, which may have altered properties due to thermal degradation, oxidation, or contamination.

In testing for ESR, materials are subjected to stress under exposure to chemicals, detergents, or oils, simulating real-world conditions. The test evaluates how quickly cracks develop and propagate, allowing for comparison between different batches of recyclates.

Since contaminants and residual stress from prior processing can increase the susceptibility of recycled materials to ESR, testing helps ensure recyclates are suitable for applications requiring high chemical resistance and durability.

3.1.9 Summary

The quality of recyclates is key to their successful use in new products. Without rigorous testing, variations in composition, contamination, and degradation can lead to inconsistent performance, limiting their applicability in high-value applications. Contamination, in particular, remains a critical challenge, as even small amounts of incompatible materials or residues can affect mechanical properties, processing behavior, and long-term durability.

By applying thermal, spectroscopic, rheological, mechanical, and structural analysis, recyclates can be qualified for use in specific applications, ensuring they meet processing and performance requirements. These tests not only help recyclers improve sorting and processing strategies but also give manufacturers confidence that recycled materials can replace virgin polymers without compromising quality.

As the demand for sustainable materials grows, proving the quality of recyclates will be essential to their acceptance in industries such as automotive, packaging, and construction. The next section provides a structured comparison of testing methods, summarizing their advantages, costs, and practical considerations to guide industry professionals in selecting the most appropriate techniques.

3.2 Comparative Overview of Testing Methods

Selecting the right testing method for recyclates depends on the intended application, the level of material characterization required, and available resources. Some techniques are quick and cost-effective for quality control (QC), while others offer detailed material insights at a higher cost and complexity.

Table 3.1 presents a practical comparison of key testing methods, outlining their applications, cost factors, required expertise, and processing times. By understanding these trade-offs, recyclers and manufacturers can implement efficient testing strategies suited to their needs.

Choosing the most suitable testing method requires balancing cost, complexity, and the level of detail needed. Routine QC relies on quick and economical methods like MFI, FTIR, or DSC, whereas advanced techniques like DMA, SEC, and HPLC provide deeper insights into mechanical stability, degradation, and chemical composition.

As regulations tighten and industry expectations rise, recyclate testing must evolve to meet higher quality standards. Standardized frameworks such as EN 18065 help define material testing requirements, ensuring consistency across the industry. The next section focuses on how EN 18065 establishes testing protocols and data quality levels (DQLs), providing a foundation for recyclate classification and certification.

Table 3.1 Comparison of Key Testing Methods for Recyclates

No.	Method	Application	Research, QC
1	DSC	Measures melting, crystallization, glass transition temperatures to identify and quantify, OIT/OOT for assessing stability	QC & research
2	TGA	Evaluates thermal stability, decomposition, and content of fillers and additives	QC & research
3	EGA	Analyzes degradation products during thermal decomposition in TGA, providing chemical analysis	Research
4	FTIR	Identifies chemical composition, often used for sorting polymers	QC & research
5	NIR	Real-time sorting of polymers in industrial recycling	QC
6	HPLC	Detects and quantifies additives, residual monomers, and degradation byproducts	Research
7	SEC	Determines molecular weight distribution, crucial for assessing degradation	Research
8	Capillary rheometry	Simulates industrial processes to assess flow properties of polymers	Research & QC
9	Rotational rheometry	Rotation: Measures flow behavior and molecular structure Oscillation: Measures viscoelastic properties as well as miscibility and compatibility of blends	Research & QC
10	MFI	Simple test to measure flow rate under a given load and temperature	QC
11	VN	Measures viscosity in solution, indirectly assessing molecular weight (changes)	QC
12	UTM	Tests tensile strength and durability of materials, creep stress relaxation	Research & QC
13	Charpy/Izod	Measures impact resistance of materials, typically for structural applications	Research & QC
14	HDT	Measures temperature at which material deforms under a load (HDT) or softens (Vicat)	Research & QC
15	DMA	Softening temperature under load, Measures elastic and viscous behavior under dynamic loading, used for creep and stress relaxation	Research
16	Light microscopy	Quickly analyzes macroscopic defects, cracks, and surface features	Research & QC
17	SEM	Provides high-resolution images of surface morphology and defects	Research
18	ESR	Evaluates resistance to cracking when exposed to mechanical stress and chemicals	Research

Instrument Price	Operating Costs	Personnel Expertise	Testing Time	Sample Preparation
Moderate	Moderate	Low to medium	2–3 hours	Simple: 5–10 mg sample from pellets/flakes
Moderate	Moderate	Low to medium	1.5–2 hours	Simple: 5–20 mg sample from pellets/flakes
High	High	High	Varies	Requires coupling with TGA
Low	Low	Low	Minutes	Minimal: pellets/flakes
Low	Low	Low	Seconds	None
High	High	High	Minutes to an hour	Requires pellet/flake sample dissolution
High	Medium	Medium	Several hours	Pellet/flake sample dissolved, filtered
High	Low	Low to medium	A few hours	Pellets/flakes
Moderate	Low	Low to medium	A few hours	Pellets/flakes
Low	Low	Low	Minutes to 0.5 hours	Pellets/fakes
Low	Low	Medium	1–2 hours	Pellets/flakes dissolved
Moderate to high	Low to moderate	Medium	Minutes (creep: days)	Requires injection molded sample
Moderate	Low to moderate	Medium	Minutes	Requires injection molded sample
Moderate	Moderate	Medium	Minutes to hours	Requires injection molded sample
Moderate to high	Low to moderate	Medium to high	Hours	Requires injection molded sample
Low	Moderate	Low	Minutes	Medium: embedding thin cuts or polishing
High	High	High	Minutes to hours	Conductive coating needed
Low to moderate	Low to moderate	Medium	Days to weeks	Requires injection molded sample

3.3 Testing Procedures for Recyclates According to EN 18065

3.3.1 Understanding EN 18065 and Data Quality Levels (DQLs)

EN 18065 (formerly DIN SPEC 91446) establishes a standardized framework for evaluating recyclates by introducing **Data Quality Levels (DQLs)**, ensuring transparent and consistent classification of materials. These levels define the depth and reliability of information available for a recyclate, guiding both producers and users in assessing material quality.

The DQL system ranges from basic to advanced testing requirements, with higher levels demanding more detailed analyses. This classification enables recyclers, converters, and manufacturers to align expectations, optimize processing conditions, and ensure compliance with industry regulations. By implementing EN 18065, companies can improve the reliability and market acceptance of recyclates, facilitating their broader use in high-performance applications. Refer to Table 3.2 for a breakdown of required tests per DQL level.

Table 3.2 Properties Specified in EN 18065, Table A2

Property	DQL 1	DQL 2	DQL 3	DQL 4
Viscosity (MVR/MFR, IV, VN)	X	X	X	X
Ash content		X	X	X
Residual moisture content		X	X	X
Density		X	X	X
Bulk density			X	X
Particle size distribution				X
Tensile properties				X
Material identification (FTIR or DSC)				X

Currently, many standards are being developed to support the classification and characterization of recyclates beyond EN 18065. This includes standards for specific polymer types, such as the DIN EN 1534X series, which covers materials like polystyrene (PS), polyethylene (PE), polypropylene (PP), polyvinyl chloride (PVC), and polyethylene terephthalate (PET). Additionally, industry-specific standards are emerging for packaging, automotive, and other sectors, further reinforcing quality expectations for recyclates in various applications.

3.3.2 Key Testing Methods for Recyclates

To meet the requirements of EN 18065, recyclates are assessed using various testing methods that evaluate processability, composition, and mechanical performance.

Viscosity is a fundamental parameter in polymer processing, indicating molecular weight and flow behavior, as explained in Section 3.1.5. The **melt flow index** (MFI or MFR) and **melt volume-flow rate** (MVR) are commonly used for routine quality control, while **intrinsic viscosity** (IV) and **viscosity number** (VN) provide more precise molecular weight assessments for PET and polyamides.

Ash content analysis identifies the presence of inorganic fillers, additives, or contaminants, which influence mechanical and thermal properties. It is measured via **thermogravimetric analysis (TGA)** (see Section 3.1.3) or **muffle furnace incineration**, ensuring recyclates meet purity requirements.

Excess moisture affects processing stability and mechanical performance, particularly for hygroscopic polymers like polyamides and polyesters. **Karl Fischer titration**, **moisture analyzers**, and TGA are used to measure and control residual moisture before processing.

Density measurements verify material consistency and detect contamination. **Pycnometry** or density set for an analytical balance (Archimedes' principle) provides precise density data, while bulk density measurements are used to assess flow behavior in processing systems, particularly for flakes or powders.

Recyclates in granulated or powdered form require consistent particle size distribution (PSD) to ensure uniform feeding and processing. **Sieving analysis** and **laser diffraction** methods determine PSD and prevent processing inconsistencies.

Mechanical performance is assessed through tensile strength, elongation at break, and Young's modulus as described in Section 3.1.6. These properties determine whether recyclates can directly replace virgin materials or require blending with virgin polymers for enhanced performance.

To confirm polymer type and detect contamination, **Fourier-transform infrared spectroscopy (FTIR)** and **differential scanning calorimetry (DSC)** are employed. FTIR (see Section 3.1.4) provides a molecular fingerprint for polymer identification, while DSC (see Section 3.1.3) determines melting and crystallization behavior, ensuring recyclate consistency.

4

Environmental and Economic Analysis of Plastics Recycling

The transformation towards a circular plastics economy requires more than just technical innovation – it demands solutions that are both environmentally sound and economically viable. As environmental concerns surrounding plastic waste continue to rise globally, the sustainability of recycling methods has become a central topic in research, policy, and industry. This chapter focuses on the environmental and economic dimensions of plastics recycling, highlighting the interdependence between ecological impact and financial feasibility.

The chapter highlights that various recycling methods exhibit differing environmental performances, particularly concerning energy consumption, greenhouse gas emissions, and resource conservation. For example, producing recycled high-density polyethylene (HDPE) and polypropylene (PP) can reduce total energy consumption by up to 88% compared to virgin resin production. Additionally, recycled resins can cut greenhouse gas emissions by approximately 67% for polyethylene terephthalate (PET) and 71% for both HDPE and PP. These reductions contribute significantly to climate goals and decrease reliance on fossil-based raw materials [1]. While HDPE and PP are already relatively energy-efficient in their virgin form, the energy required to produce different plastics can vary greatly. This makes recycling an especially valuable tool – not only for resource-intensive engineering plastics, but also for commonly used polyolefins.

On the economic front, the global recycled plastics market was valued at over USD 51 billion in 2023 and is projected to reach USD 107 billion by 2032, reflecting a compound annual growth rate (CAGR) of 8.6%. However, profitability varies across recycling methods and material streams. The book examines key profitability drivers, including input quality, process efficiency, and market dynamics, providing evaluations of different recycling technologies based on their economic potential [2].

By presenting a holistic view, the chapter provides essential insights into how environmentally and economically balanced recycling solutions can be designed and implemented to support a sustainable, circular plastics economy.

4.1 Influence Factors of Environmental Impact of Plastics Recycling

The environmental performance of plastics recycling is influenced by a wide range of interrelated factors, spanning from energy and material efficiency to pollution potential and infrastructure capabilities. This section outlines and explores the key aspects that shape the ecological footprint of recycling processes, providing a holistic view of their environmental impact. Thereby, these factors are partially interlinked. Table 4.1 summarizes the factors which will be analyzed in the following sub-chapters. Section 4.1.7 summarizes the factors and compares them for both mechanical and chemical recycling.

Table 4.1 Influence Factors of Environmental Impact of Plastics Recycling

Factor	Description	Relevance
Energy Consumption | Section 4.1.1	Total energy required for collection, sorting, processing, and reprocessing.	Strongly affects operational efficiency and carbon footprint.
Carbon Footprint | Section 4.1.2	All greenhouse gas emissions across the recycling life cycle.	Core metric for climate impact assessment and environmental sustainability.
Input Material Quality & Complexity | Section 4.1.3	Refers to the type, purity, and design of plastic waste (e. g., multilayers, additives).	Determines process compatibility, recyclate quality, and efficiency.
Process Emissions & Pollution | Section 4.1.4	Non-GHG emissions such as VOCs, microplastics, toxic byproducts, and wastewater.	Impacts local environmental and health risks; varies by technology.
Waste Management Infrastructure | Section 4.1.5	Availability and efficiency of collection, sorting, and regulatory systems.	Key enabler for high recycling rates and low contamination.
Contribution to the Circular Economy | Section 4.1.6	The extent to which recycling keeps materials in high-value use.	Indicates whether recycling supports long-term resource loops or downcycling.

4.1.1 Energy Consumption

Energy consumption is a key indicator in assessing the environmental performance of plastics recycling. It not only reflects the efficiency of resource use but also directly influences greenhouse gas emissions (cf. Section 4.1.2) Understanding energy demand is essential when comparing different recycling technologies and evaluating their sustainability.

Several factors determine the total energy consumption of a recycling process:

- **Polymer type:** The melting point and structural complexity of polymers affect energy requirements. For instance, PET requires more thermal energy than polyolefins (cf. Section 4.1.3).

- **Input contamination:** Heavily contaminated plastics require intensive pre-treatment, including multiple washing and drying stages, which increase energy use (cf. Section 4.1.3).

- **Process design and integration:** Plants that use heat recovery systems or closed-loop water circuits are generally more energy-efficient.

- **Energy source:** Whether energy is derived from fossil fuels or renewable sources significantly influences both energy efficiency and environmental impact.

Mechanical recycling is widely regarded as the most energy-efficient method for processing plastic waste. It involves sorting, washing, shredding, drying, and remelting. The energy consumption of mechanical recycling typically ranges from 0.4 to 1.5 MJ/kg of plastic, depending on the polymer and the system design [3]. Compared to virgin plastic production, mechanical recycling yields substantial energy savings. For example, producing recycled PET requires approximately 79% less energy than producing virgin PET from petrochemical feedstocks [4]. However, inefficiencies due to input contamination or outdated equipment can reduce these benefits.

Chemical recycling technologies, such as pyrolysis, gasification, and depolymerization, are more energy-intensive due to the high temperatures and reaction kinetics involved. Reported values range from 5 to 15 MJ/kg per input material [5]. The wide variation stems from differences in process design, polymer type, and plant efficiency. While chemical recycling enables the processing of plastics that cannot be mechanically recycled – such as multilayer films or heavily contaminated materials – it requires significant thermal or chemical energy inputs. Integration with renewable energy or energy recovery systems is crucial to improve the sustainability profile of these technologies.

4.1.2 Carbon Footprint (Greenhouse Gas Emissions)

The carbon footprint of plastics recycling refers to the total greenhouse gas emissions associated with the process, typically expressed as CO_2-equivalents per kilogram of material processed. While energy consumption (cf. Section 4.1.1) is the dominant factor, other sources of emissions also contribute and must be considered for a complete life cycle perspective.

- **Process-specific emissions:** In chemical recycling, thermal degradation and catalytic reactions can produce additional greenhouse gases, including methane, carbon monoxide, and volatile organic compounds. These emissions depend on process control, reactor design, and whether emissions are captured or flared [5].

- **Energy source and intensity:** The carbon intensity of the energy used in recycling processes greatly influences emissions. For instance, 1 kWh of electricity from coal corresponds to an emission of approximately 1 kg of CO_2, while the same energy from solar or wind produces close to zero direct emissions. Therefore, two identical recycling plants powered by different energy sources can have vastly different carbon footprints [6].

- **Ancillary materials and water treatment:** Emissions from the production and use of cleaning agents, catalysts, and water treatment chemicals – often used in washing or solvent-based processes – also contribute to the total footprint. Heated water circuits further increase energy demand [3].

- **Transportation and logistics:** The collection and transport of plastic waste contribute to indirect emissions, especially when waste is shipped over long distances or exported. Localized recycling facilities significantly reduce transport-related emissions [7].

- **Residue management and end-of-life:** Not all material entering a recycling process is recovered. Rejects, filters, and non-recyclable fractions are often incinerated or landfilled. These disposal routes may contribute up to 30% of the total GHG emissions in some mechanical recycling systems [3].

Mechanical recycling typically results in a 50–70% reduction in carbon emissions compared to virgin plastic production, depending on polymer type and system efficiency [6]. This is due to the lower energy requirements and minimal chemical transformations involved.

In contrast, chemical recycling can have a comparable or even higher carbon footprint than virgin production, especially if powered by fossil fuels or if process yields are low. However, when integrated with renewable energy, high-efficiency systems, or carbon capture technologies, the carbon impact of chemical recycling can be significantly reduced [5].

It is necessary to note that the interpretation of carbon footprints depends heavily on the system boundaries used in life cycle assessments (LCA):

- **Cradle-to-gate** studies cover emissions up to the point the recyclate is produced.

- **Cradle-to-grave** studies include product use and end-of-life.

- **Cradle-to-cradle** scenarios include closed-loop recycling, offering a more complete view of long-term environmental benefits.

LCA is an essential tool to assess the net carbon benefit of recycling systems and to identify emission hotspots across the entire value chain.

4.1.3 Input Material Quality and Complexity

The type and quality of plastic waste are crucial determinants of recycling efficiency and environmental impact. Plastics vary in recyclability based on their polymer type, degree of contamination, use of additives, and structural complexity. Low-quality or complex inputs require more energy and processing and often yield lower-quality recyclates or generate more waste.

Mechanical recycling is highly sensitive to input quality. Clean, homogenous streams result in high material yields and low environmental impact. In contrast, contaminated or mixed plastic streams require extensive sorting and washing, increasing energy consumption (cf. Section 4.1.1) and lowering process efficiency. Additives such as pigments or flame retardants can negatively affect the quality of the recycled material, and multilayer or composite plastics are often non-recyclable through mechanical methods [3].

Chemical recycling can handle more complex or contaminated plastics than mechanical systems. It is capable of processing mixed polymers and multilayer materials. However, the presence of chlorine, fluorine, or heavy metals can damage catalysts or reactors, and high feedstock variability may reduce product quality and process stability. Cleaner input still improves overall performance [5].

4.1.4 Process Emissions and Pollution

Beyond carbon emissions, recycling processes can generate other pollutants, including airborne emissions, wastewater, toxic byproducts, and microplastics. These impacts depend on process type, input materials, and emission control systems.

The key environmental risks in mechanical recycling are:

- **Microplastic generation** during shredding and washing stages
- **Wastewater pollution** from detergents and organics unless closed-loop water systems are used
- **Emissions** from thermal energy use in drying and extrusion
- **Toxic byproducts** if input materials are not controlled

These impacts can be managed with good operational practices and infrastructure but remain important environmental considerations [8].

The key environmental risks in chemical recycling are:

- **Air pollutants** (e. g., NO_x, VOCs, dioxins) during high-temperature operations such as pyrolysis and gasification
- **Toxic byproducts**, especially when halogenated plastics (e. g., PVC) are present
- **Solid residues** (char, ash) that may be hazardous and require regulated disposal

Effective emissions mitigation requires flue gas treatment, advanced reactors, and input control. While pollution can be controlled, it poses a greater environmental risk than in mechanical systems [5].

4.1.5 Waste Management and Infrastructure

Recycling systems do not exist in a vacuum – they depend on infrastructure such as collection, sorting, regulation, and public participation. Strong infrastructure leads to higher recycling rates, lower contamination, and improved environmental outcomes.

Mechanical recycling relies heavily on well-developed infrastructure. Efficient collection and source separation, high-tech sorting facilities, and clear product labeling are critical to avoid contamination and improve output quality. Where these systems are absent, performance declines significantly [9].

Chemical recycling can complement mechanical systems in regions with limited infrastructure. However, it still requires reliable and consistent feedstock supply, regulatory approval, and capital investment. While it has more flexibility in input material, it remains dependent on broader waste management systems with collection as the first step [10].

Global disparities in waste management infrastructure are striking. Germany leads with a municipal recycling rate of around 66%, supported by decades of investment in public education, separate collection systems, and policy enforcement [11]. In contrast, the United States recycles only about 32% of its municipal waste, largely due to inconsistent local systems and limited nationwide mandates [12]. China has improved in recent years but still recycles only about 24% [13], while India faces greater challenges with a municipal recycling rate of just 18% [14]. In many regions, especially in developing countries, waste collection infrastructure is still lacking – making any form of recycling difficult to scale.

These differences highlight the importance of infrastructure investment and policy development as foundational steps toward a functioning circular economy.

4.1.6 Contribution to the Circular Economy

Recycling contributes to the 9R of circular economy by keeping materials in use, reducing the need for virgin plastic, and lowering environmental impact. Its value depends on how well it preserves material quality and enables closed-loop applications (cf. Chapter 1).

Mechanical recycling supports closed-loop systems for polymers like PET and HDPE, enabling applications such as bottle-to-bottle recycling. However, thermal and mechanical degradation over multiple cycles often leads to downcycling (see Chapter 2), where the recycled material is used for lower-value products [5, 8].

Chemical recycling can regenerate virgin-quality feedstock, such as monomers or fuel components. This offers the potential to recycle contaminated or complex plastics into high-value applications. However, its contribution to circularity depends on the energy source and life cycle emissions [15].

4.1.7 Summary of the Environmental Influence Factors for Mechanical and Chemical Recycling

Table 4.2 Comparison of Environmental Factors of Mechanical and Chemical Recycling

Factor		Mechanical Recycling	Chemical Recycling
Energy consumption	Typical energy use	0.4–1.5 MJ/kg	5–15 MJ/kg
	Key factors	Contamination, water use, polymer type	Technology type, temperature, feedstock
Carbon footprint	Average CO_2 footprint	0.4–0.8 kg CO_2-eq/kg	1.5–3.0 kg CO_2-eq/kg
	Key contributors	Electricity mix, input contamination, reject disposal	High-temperature energy use, process emissions, fossil input
Input material quality and complexity	Sensitivity to input quality	High	Moderate
	Processing complex plastics (mixtures)	No	Yes
Process emissions and pollution	Air pollution	Low	High (if untreated)
	Microplastics	Yes	No
	Toxic byproducts	Minimal	High (if not controlled)
Waste management infrastructure	Infrastructure dependency	Very high	High
	Main challenges	Sorting, contamination, standardization	Feedstock control
Contribution to circular economy	Closed-loop potential	High (for clean streams)	High (monomer recovery)
	Downcycling risk	Moderate–high	Low
	Circularity potential	Strong for single-polymer waste	High (if energy is renewable)

Table 4.2 summarizes the main environmental influence factors for both mechanical and chemical recycling. This shows that neither one of the methods is more sustainable than the other, but that it depends on the considered factors. Mechanical recycling on the one hand requires less energy and has a lower carbon footprint. Chemical recycling on the other hand can process a lot more plastics and has a higher contribution to circular economy.

4.2 Influence Factors of Economics of Plastics

This chapter analyzes the key economic drivers that affect the viability and profitability of plastics recycling systems. Each section introduces a relevant factor, and then compares its impact on mechanical and chemical recycling. Table 4.3 summarizes the factors which will be analyzed in the following sub-chapters. Section 4.2.7 summarizes the factors and compares them for both mechanical and chemical recycling.

Table 4.3 Influence Factors of Economic Impact of Plastics Recycling

Factor	Description	Relevance
Material Supply and Feedstock Quality | Section 4.2.1	Availability, collection, consistency, and cleanliness of plastic input materials.	Determines operational efficiency, cost of pretreatment, and quality of recyclate.
Operational Efficiency and Economies of Scale | Section 4.2.2	Process performance, automation level, and facility size.	Drives per-unit cost reduction and improves profitability through optimized operations.
Market Demand and Price Volatility | Section 4.2.3	Market prices for recycled vs. virgin plastics and demand for recycled content.	Affects revenue and competitiveness; volatile prices can destabilize business models.
Regulatory and Policy Incentives | Section 4.2.4	Laws, mandates, taxes, and subsidies related to recycling and circular economy.	Influences market creation, investment attractiveness, and system viability.
Costs and Capital Investment | Section 4.2.5	Operational expenses and upfront capital needed to establish recycling systems.	Defines entry barriers and long-term profitability across technologies.
Market Access, Trade, and Logistics | Section 4.2.6	Access to markets, export/import regulations, and transportation infrastructure.	Shapes collection and distribution costs and the ability to sell recycled materials profitably.

4.2.1 Material Supply and Feedstock Quality

A reliable and high-quality input stream is foundational to the economic performance of any plastics recycling operation. Material supply involves not only the availability through collection or the volume of incoming waste plastics, but also their consistency, polymer type, and degree of contamination. Variations in these characteristics significantly influence operational costs, recovery efficiency, and the market value of the final recyclate.

In mechanical recycling, profitability is highly dependent on the availability of clean, homogenous, and easily separable plastic waste. High-volume postconsumer waste streams such as PET and HDPE bottles are particularly attractive due to their established collection systems and favorable material properties. However, contamination – whether from food residues, labels, or non-target polymers – increases the cost of sorting, washing, and reprocessing, often resulting in lower-quality recyclates or rejected materials. Complex structures, such as multilayer films or blends, pose additional challenges and are generally uneconomical to recycle via mechanical methods [3].

Chemical recycling provides greater flexibility in handling diverse and contaminated inputs. Technologies such as pyrolysis and gasification can process mixed polyolefins and contaminated waste streams that are otherwise unrecyclable. Nonetheless, certain contaminants – especially halogenated plastics like PVC – must be excluded to prevent equipment corrosion and hazardous emissions. Moreover, feedstock variability can reduce product quality and yield, thereby impacting profitability. Effective feedstock preparation and consistent quality remain important for optimizing output in chemical recycling processes [5].

4.2.2 Operational Efficiency and Economies of Scale

Operational efficiency and plant scale are key determinants of recycling costs and margins. Efficient use of labor, energy, water, and equipment allows recyclers to lower their per-unit costs and increase profitability. Economies of scale, achieved through large-scale processing and integrated collection networks, further enhance economic performance.

Mechanical recycling plants benefit significantly from automation in sorting, shredding, and extrusion. Optical sorters, robotics, and AI-based quality control reduce labor costs and increase throughput (see Chapter 2). Moreover, facilities with integrated water and heat recovery systems achieve lower operational expenses. Larger-scale plants can amortize equipment costs over greater volumes, improving economic viability [16].

Chemical recycling operations are generally more capital-intensive and less mature, but similar principles apply. High-temperature reactors and chemical conversion units benefit from economies of scale and process integration. However, energy use is higher than in mechanical systems, making energy efficiency and plant optimization even more critical. Modular, scalable reactor designs and integrated utility systems are emerging trends aimed at improving economic returns [5, 17].

The location of recycling facilities plays a key role in achieving these efficiencies. Larger urban areas – with their high population density and consistent material flow – favorable conditions for economies of scale. Centralized plants in metropolitan regions can operate continuously at high capacity, spreading fixed costs over greater volumes and enabling advanced process integration. In contrast, rural areas face challenges such as lower waste generation volumes, greater distances between collection points, and limited infrastructure. These factors increase transportation and operating costs, making it more difficult to justify capital-intensive investments. For example, one study found that collection costs in low-density areas can be up to twice as high as those in urban environments, primarily due to greater transportation distances and lower route efficiency [18].

4.2.3 Market Demand and Price Volatility

The financial viability of recycled plastics depends heavily on market demand and price dynamics. Recycled resin prices are influenced by supply and demand, virgin material prices (linked to oil markets), product quality, and industry adoption of recycled content.

Oil is the most important raw material for plastics. 1 kg of plastic requires about 2 kg of crude oil (including processing and raw materials for plastics). For this reason, the oil price has a great influence on the plastic price and the profitability of the whole plastic recycling process. The lower the oil price, the lower the price of the recycled plastic, the less profitable the recycling process becomes. Furthermore, a lower oil price reduces the costs of producing new, or virgin, plastic material, which is another challenge for plastics recycling.

In 2015 and 2016, low global oil prices significantly reduced the profitability of plastics recycling in the United States. In Newark, for example, the value for 1 bale of recycled plastics decreased from $230 to $112. One of the consequences was that Infinitus Energy, which had just opened a $35 million recycling center in Montgomery, Alabama, in 2014, shut that facility down in October 2014 since it was incurring losses only [19].

Figure 4.1 shows the changing price of 1 t of regrind PET compared to 1 barrel (158.9873 L) of oil. Since the original prices of the plastics were in Euros, the rate of 1 Dollar per Euro was taken from X-Rates. For a better comparison of different years, prices were inflation-adjusted based on the prices of October 2008 (both PET and oil) [20, 21, 22, 23].

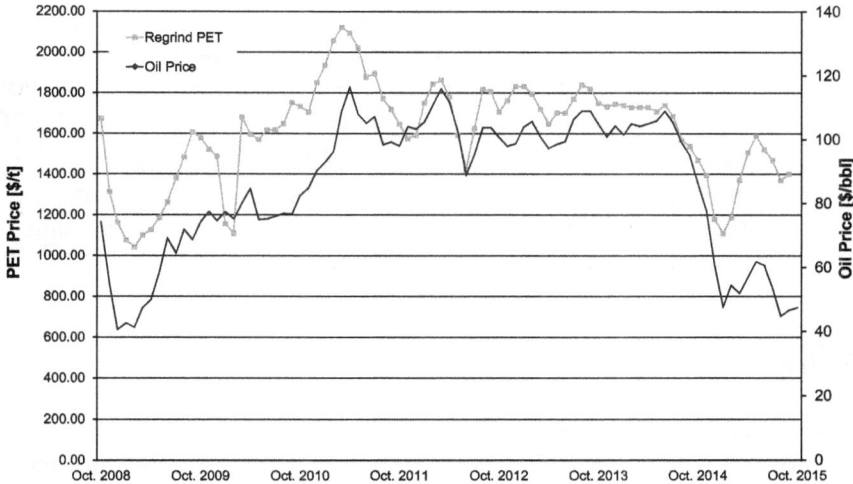

Figure 4.1 Oil (1 barrel (bbl) = 158.9873 L) and PET (1 t regrind) prices between October 2008 and October 2015 (inflation-adjusted)

Figure 4.1 demonstrates that the PET price is following the oil price changes, since the peaks of the regrind PET price graph line are lagging the peaks of the oil barrel price graph line by approximately 1 month. This can especially be observed in November 2009, October 2010, November 2014, and August 2015.

In mechanical recycling, profitability is closely tied to stable demand for high-quality recyclates such as rPET and rHDPE. When oil prices are low, virgin plastics become cheaper, exerting downward pressure on recycled resin prices. This often makes mechanical recyclers vulnerable to market shifts unless supported by regulations or brand commitments to recycled content. High-grade, food-contact-compliant materials typically fetch better prices and ensure greater financial resilience [16, 24].

Chemical recycling has the potential to produce high-purity, virgin-equivalent outputs. This can command premium prices, especially in sectors with strict regulatory requirements (e. g., food packaging). However, most chemical recycling markets are still nascent, and pricing structures remain volatile. Market acceptance, certification standards, and brand uptake of chemically recycled materials will determine long-term profitability [25].

4.2.4 Regulatory and Policy Incentives

Public policy is increasingly shaping the economics of plastics recycling through regulations, mandates, and economic instruments. Key levers include Extended Producer Responsibility (EPR), recycled content targets, tax incentives, and carbon pricing mechanisms. This will be further analyzed in Chapter 5.

Mechanical recycling benefits greatly from EPR schemes and recycled content mandates, which increase demand for recycled resins and ensure financing for collection and sorting infrastructure. EU policies, for example, are creating more stable markets for recyclates by enforcing packaging targets and product labeling requirements [9, 26].

Chemical recycling also stands to benefit from policy innovation. However, its recognition under circular economy and EPR frameworks remains inconsistent across jurisdictions. Where recognized, chemical recycling can benefit from technology subsidies, circular economy funding, and carbon credit schemes for avoided emissions [27].

4.2.5 Operating Costs and Capital Investment

Economic performance hinges on both ongoing operational costs and the upfront capital required to establish a recycling facility. Key costs include labor, maintenance, electricity, water, and transport. Capital investment varies widely between technologies.

Mechanical recycling facilities generally require moderate capital investment for shredders, sorters, washers, and extruders. Costs scale with capacity but are usually lower than those for chemical recycling. Operating costs are heavily influenced by energy and water use, input contamination, and maintenance requirements [3].

Chemical recycling involves substantially higher capital expenditure, particularly for pyrolysis, gasification, or depolymerization systems. These require reactors, condensers, feedstock treatment systems, and advanced emission controls. Operating costs are high due to energy intensity, chemical consumption, and residue management. As a result, profitability depends on plant scale, output quality, and revenue from by-products or fuels [5, 17].

4.2.6 Market Access, Trade, and Logistics

The geographical location of recycling facilities and the structure of global recycling markets affect profitability through logistics and market access. Transportation costs, export restrictions, and local market saturation all influence economic outcomes.

Mechanical recycling is most profitable when plants are located close to both waste sources and end-markets. Long-distance transportation of low-value plastic waste adds significant cost and emissions. Moreover, international market disruptions, such as China's National Sword policy (cf. Chapter 5 and Chapter 6), have limited export routes for low-grade recyclates, pushing recyclers to develop domestic markets [9, 28].

Chemical recycling may be more viable in centralized facilities processing mixed or difficult waste streams, particularly when integrated with industrial clusters. However, transport of heterogeneous waste and refined outputs still contributes to total system costs. Ensuring regional supply security and infrastructure integration is crucial for long-term success [25].

4.2.7 Summary of the Economic Influence Factors for Mechanical and Chemical Recycling

Table 4.4 summarizes the main economic influence factors for both mechanical and chemical recycling.

Table 4.4 Comparison of Economic Factors of Mechanical and Chemical Recycling

Factor		Mechanical Recycling	Chemical Recycling
Material supply and feedstock quality	Sensitivity to input	Requires clean, sorted, mono-material streams	Tolerates mixed and contaminated inputs
	Contamination tolerance	Low	Moderate
	Resin type preference	PET, HDPE, PP (high value, high volume)	Polyolefins, mixed streams
Operational efficiency and economies of scale	Automation potential	High (e. g., optical sorters, robotics)	Medium (under development)
	Capital efficiency	Better at small-to-medium scale than chemical recycling, but large is preferred	Requires large scale for profitability
	Efficiency drivers	Sorting, cleaning, extrusion	Reactor design, integration, energy optimization
Market demand and price volatility	Price stability	Vulnerable to virgin plastic price fluctuations	Premium potential, but currently volatile
	Product quality	Variable; downcycling risk	Virgin-equivalent output possible
	Market maturity	High (e. g., rPET, rHDPE, rPP)	Emerging
	Key challenges	Competing with low-cost virgin plastics	Gaining market trust and certification
Regulatory & policy incentives	EPR & content mandates	Strong positive influence	Depends on recognition under EPR
	Access to subsidies	Widely supported (infrastructure, CAPEX)	Supported in some regions; not universal
	Carbon credit relevance	Emerging	High potential (if integrated with circular/ energy goals)

Table 4.4 Comparison of Economic Factors of Mechanical and Chemical Recycling
(continued)

Factor		Mechanical Recycling	Chemical Recycling
Operating costs and capital investments	Capital intensity	Moderate	High
	Operating costs	Sensitive to energy, water, and labor	High (energy, chemical inputs, residue management)
	Maintenance requirements	Regular (medium complexity systems)	Complex (reactors, flue gas treatment)
	Scale sensitivity	Scales well from medium-sized plants	Requires industrial scale to be competitive
Market access, trade & logistics	Location dependency	High – proximity to feedstock and markets matters	Moderate – centralization possible but still logistically complex
	Export vulnerability	High (e. g., China's National Sword policy impact)	Lower – less dependent on global waste trade
	Integration advantage	With local waste systems	With industrial parks, energy networks

4.3 Comparative Environmental and Economic Analysis of Mechanical and Chemical Plastics Recycling

This section summarizes the environmental and economic insights discussed in previous chapters by comparing mechanical and chemical recycling across key performance dimensions.

Across both systems, key shared variables include feedstock quality, process efficiency, and energy consumption. Table 4.5 summarizes the major cross-cutting factors influencing the environmental and economic results.

Table 4.5 Cross-Cutting Factors Influencing Environmental and Economic Outcomes

Factor	Environmental Relevance	Economic Relevance
Feedstock quality	Reduces emissions, contamination, and microplastics	Improves yield, lowers sorting and cleaning costs
Material homogeneity	Enables closed-loop recycling and high-quality outputs	Supports higher market prices and reduces reprocessing needs
Energy consumption	Major driver of carbon footprint	Key component of operational expenditure
Process efficiency	Reduces emissions and material losses	Enhances throughput and cost-effectiveness
Regulatory environment	Drives sustainability through mandates and incentives	Creates stable markets and improves investment security

Besides these cross-cutting factors influencing the environmental and economic outcomes, the chapter highlighted six key decision factors for mechanical and chemical recycling, as Table 4.6 summarizes. For mechanical recycling, success is contingent on the availability of clean and sorted materials – reducing both emissions and costs. For chemical recycling, pre-treatment of complex feedstock and process optimization are crucial for achieving marketable outputs while minimizing environmental burdens.

Table 4.6 Key Decision Factors for Mechanical and Chemical Recycling

Criteria	Mechanical Recycling	Chemical Recycling
Energy and emissions	Low energy use, low GHG emissions	High energy use, higher GHG emissions (if fossil-based)
Feedstock requirements	Requires clean, mono-material streams	Accepts complex/mixed plastics, but still sensitive to PVC
Product quality	Risk of degradation, suitable for lower end uses	Virgin-like quality, suitable for demanding applications
Capital and operating costs	Moderate CAPEX and OPEX	High CAPEX and OPEX
Market and policy maturity	Mature markets, policy support well established	Emerging markets, policy recognition developing
Circularity potential	Strong for simple polymers (e. g., PET)	High if powered by renewable energy and integrated well

A concrete economic analysis for recycling of PET with mechanical recycling is in the appendix of this book (Chapter 8). Here, a detailed analysis of the different process steps, including a dynamic calculation sheet, is provided. Furthermore, different optimization scenarios are analyzed.

References

[1] Pyzyk, K. *APR: Recycled plastics reduce energy consumption, GHG emissions*. Access Date: 19.04.2025. Available: *https://www.wastedive.com/news/apr-recycled-plastics-reduce-energy-consumption-ghg-emissions/547027/?utm*.

[2] Fortune Business Insights. *Recycled Plastics Market Size, Share & Industry Analysis*. Access Date: 18.04.2025. Available: *https://www.fortunebusinessinsights.com/recycled-plastic-market-102568*.

[3] Ragaert, K., Delva, L., and Van Geem, K. Mechanical and chemical recycling of solid plastic waste. *Waste Management*. vol. 69. 2017. pp. 24–58.

[4] Hopewell, J., Dvorak, R., and Kosior, E. Plastics recycling: challenges and opportunities. *Philosophical Transactions of the Royal Society B: Biological Sciences*. vol. 364. 2009. pp. 2115–2126.

[5] Jeswani, H., Krüger, C., Russ, M., Horlacher, M., Antony, F., Hann, S., and Azapagic, A. Life cycle environmental impacts of chemical recycling via pyrolysis of mixed plastic waste in comparison with mechanical recycling and energy recovery. *Science of the Total Environment*. vol. 769. 2021. p. 144483.

[6] Jiang, X. and Bateer, B. A systematic review of plastic recycling: technology, environmental impact and economic evaluation. *Waste Management & Research*. 2025.

[7] Rigamonti, L., Grosso, M., and Sunseri, M. C. Influence of assumptions about selection and recycling efficiencies on the LCA of integrated waste management systems. *The International Journal of Life Cycle Assessment*. vol. 14. 2009. pp. 411–419.

[8] Brown, E., MacDonald, A., Allen, S., and Allen, D. The potential for a plastic recycling facility to release microplastic pollution and possible filtration remediation effectiveness. *Journal of Hazardous Materials Advances*. vol. 10. 2023. p. 100309.

[9] OECD. *Global Plastics Outlook: Policy Scenarios to 2060*. Paris, France. 2022.

[10] Hopewell, J., Dvorak, R., and Kosior, E. *Plastics Recycling: Challenges and Opportunities*. 2009.

[11] Igini, M. How Waste Management in Germany Is Changing the Game. Access Date: 24.04.2025. Available: *https://earth.org/waste-management-germany/*.

[12] United States Environmental Protection Agency (EPA). *National Overview: Facts and Figures on Materials, Wastes and Recycling*. Access Date: 24.04.2025. Available: *https://www.epa.gov/facts-and-figures-about-materials-waste-and-recycling/national-overview-facts-and-figures-materials?utm_source=chatgpt.com*.

[13] Wu, Y. *New Business Prospects in China's Waste Recycling Market*. Access Date: 24.04.2025. Available: *https://www.china-briefing.com/news/new-business-prospects-in-chinas-waste-recycling-market/*.

[14] EARTH5R. *Waste Management in India: Challenges, Innovations, and Earth5R Case Studies*. Access Date: 24.04.2025. Available: *https://earth5r.org/waste-management-india-solutions/*.

[15] Al-Salem, S., Lettieri, P., and Baeyens, J. Recycling and recovery routes of plastic solid waste (PSW): A review. *Waste Management*. vol. 29. 2009. pp. 2625–2643.

[16] PlasticsEurope. *Plastics – the Facts 2020: An analysis of European plastics production, demand and waste data*. Brussels, Belgium. 2021.

[17] Hann, S. and Connock, T. Chemical recycling: State of play. In: Report for CHEM Trust, Eunomia. 2020.

[18] Bisaschi, L., Romano, F., Carlberg, M., Carneiro, J., Ceccanti, D., Calofir, L., and Skinner, I. Transport infrastructure in low-density and depopulating areas (Issue February). In: European Parliament: Policy Department for Structural and Cohesion Policies, Brussels. 2021.

[19] Gelles, D. Losing a Profit Motive. *New York Times*. ed. New York. 2016, p. B5.

[20] Plasticker. The Home of Plastics: Raw Materials & Prices. Access Date: 2016/04/13. Available: *http://plasticker.de/preise/preise_monat_multi_en.php*.

[21] X-Rates. US Dollar per 1 Euro Monthly average. Access Date: 21.04.2025. Available: *http://www.x-rates.com/average/?from=EUR&to=USD&amount=1&year=2008*.

[22] US Inflation Calculator. Historical Inflation Rates: 1914–2016. Access Date: 21.04.2025. Available: *http://www.usinflationcalculator.com/inflation/historical-inflation-rates/*.

[23] Trading Economics. Crude Oil. Access Date: 21.04.2025. Available: *https://tradingeconomics.com/ commodity/crude-oil*.

[24] WRAP. *Plastics Market Situation Report 2021*. 2021.

[25] Ellen Macarthur Foundation. *The Global Commitment 2022 Progress Report*. 2022.

[26] European Commission. *A New Circular Economy Action Plan*. Brussels. 2020.

[27] Liu, Q., Martinez-Villarreal, S., Wang, S., Tien, N. N. T., Kammoun, M., De Roover, Q., Len, C., and Richel, A. The role of plastic chemical recycling processes in a circular economy context. *Chem. Eng. J*. 2024. p. 155227.

[28] Brooks, A. L., Wang, S., and Jambeck, J. R. The Chinese import ban and its impact on global plastic waste trade. *Science Advances*. vol. 4. 2018. eaat0131.

5

Policy and Regulations of Plastics Recycling

Policies and regulations are central levers in shaping the development and effectiveness of plastic recycling systems. They define the legal frameworks within which recycling occurs, incentivize or mandate behaviors, and influence the market dynamics between virgin and recycled plastics. Historically, regulatory responses to plastic waste have been reactive, with a focus on waste collection and disposal. However, recent years have seen a shift toward circular economy principles, emphasizing prevention, design, reuse, and material recovery. This shift is driven by growing environmental concerns, public pressure, and the need for resource efficiency. As governments recognize the scale and complexity of plastic pollution, regulatory approaches have broadened to address the full life cycle of plastic products – from production and consumption to post-use recovery and end-of-life treatment.

5.1 Policy Instruments for Promoting Recycling

Effective policy frameworks typically involve a combination of instruments that regulate market behavior, incentivize sustainable practices, and raise awareness. These instruments can be grouped into four main categories: regulatory (Section 5.1.1), economic (Section 5.1.2), informational (Section 5.1.3), and technology specific (Section 5.1.4) [1]. Figure 5.1 summarizes these four categories and their main examples.

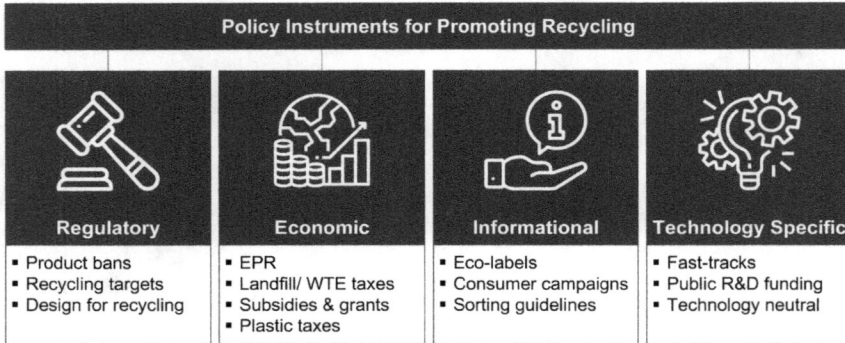

Figure 5.1 Overview of the four main categories of policy instruments for promoting recycling

5.1.1 Regulatory Instruments

Regulatory instruments form the legal backbone of recycling systems. They are designed to ensure compliance through mandates and prohibitions. These instruments can compel producers and consumers to change their behavior and support system-level change.

Examples include:

- **Product bans:** Many countries have banned specific single-use plastic items, such as straws, bags, and cutlery, to reduce litter and marine pollution [2].

- **Recycling targets:** The EU's Packaging and Packaging Waste Directive requires member states to achieve at least 55% recycling of plastic packaging by 2030 [3].

- **Design for recycling requirements:** Policies increasingly demand that packaging be compatible with existing recycling infrastructure. For instance, the EU mandates that bottle caps remain attached to bottles to reduce cap litter [4].

These measures are essential for setting clear expectations and providing long-term predictability for industry. However, the regulations must be realistic. For example, a recent study of conversion predicts, that Germany cannot comply with the laws by 2030 due to a lack of available recycled plastic material [5].

5.1.2 Economic Instruments

Economic instruments create financial incentives or disincentives that influence the behavior of producers and consumers. They are powerful tools for internalizing environmental externalities and aligning private actions with public goals [1, 6].

Common instruments include:

- **Extended Producer Responsibility (EPR):** Under EPR, producers are responsible for the costs of collecting, sorting, and recycling their products. EPR schemes often fund recycling systems and encourage product redesign [7].

- **Landfill and incineration taxes:** By increasing the cost of disposal, these taxes make recycling more attractive [6].

- **Subsidies and grants:** Financial support for recycling infrastructure and innovation can accelerate deployment and scale [8].

- **Plastic taxes:** For example, the EU charges a levy of €800 per ton of non-recycled plastic packaging waste [9].

Table 5.1 Comparison of Economic Instruments across Selected Countries

Country	EPR	Plastic Tax	Landfill Tax	Subsidies for Recycling
Germany	Yes	No	High	Yes
United Kingdom	Yes	Yes (£200/t)	Moderate	Yes
United States	Partial (state-level)	No	Varies by state	Some federal grants
Japan	Yes	No	Moderate	Yes
India	Yes	Planned	Low	Yes

5.1.3 Informational Instruments

Informational instruments aim to change behavior by raising awareness, providing transparency, and educating stakeholders. While not mandatory, they can complement other instruments by guiding voluntary action [2].

Examples include:

- **Eco-labels:** Labels such as the Blue Angel (Germany) or EU Ecolabel indicate that a product meets environmental criteria, including recyclability [10].

- **Consumer awareness campaigns:** Campaigns can shift public behavior toward better sorting, reuse, and recycling [11].

- **Sorting guidelines:** Clear, harmonized guidance helps reduce contamination in recycling streams and improves material recovery [12].

These instruments are particularly effective when embedded within broader policy frameworks. Even though the topics of recycling, single-use packaging, and plastic waste in the environment have moved into the public eye, not all efforts to reduce and recycle waste on the individual level have the same success rate.

Different studies show that fines lead to more environmentally friendly behavior than incentives. For example, a fee of 5 cents for the use of disposable bags resulted in 45% of shoppers bringing their own reusable grocery bag, while the same 5 cents bonus generated no effect on the consumers' behavior [13]. Various studies by coffee shops on reusable cup usage have shown the same [14, 15, 16]. Consumers are more motivated to avoid a fine, which is known as the model of loss aversion in behavioral economics. Early studies by Kahneman and Tversky [17] suggest that individuals perceive losses more strongly than gains. In their experiments, they posed two different problems to the individuals. One was phrased as a gain and one as a loss. In problem A they were given $1,000 and two options: a 50% chance to win another $1,000 or a gift of $500. Most individuals picked the certain gain, the gift of $500. In problem B they were given $2,000 and two options that were phrased as losses: a 50% chance to lose $1,000 or a certain loss of $500. Almost everyone now picked the gamble to potentially lose $1,000. The results of both problems were the same. In both cases the gamble gave people the 50% chance of getting $2,000. The certain option would give them $1,500. However, they perceived the pain to lose money more strongly than the pleasure to get more.

Four recent studies [16] further suggest that the framing as a fine versus an incentive signals people a social norm for this behavior. Two different scenarios – one with reusable shopping bags and one with reusable coffee mugs – were studied separately. In both cases, one group was told that a charge was imposed for not bringing a reusable cup or bag, while the other group was told that a discount was offered for bringing it. The first group inferred that bringing your item was more common and even more socially expected. The first groups commented further that they would feel guiltier for not bringing their own cup or bag.

In another of these four studies, the long-term effect of surcharges was researched. It was found that fines have a more lasting effect than incentives and can therefore lead to lasting behavioral changes.

Instead of monetary fines or incentives, another research team [18] studied the effect of positive feedback on behavioral change. In a hospital the medical staff was asked to increase the frequency of hand sanitization to prevent the spreading of diseases. Even though the importance was stressed frequently and the staff was knowingly recorded, only 10% of the medical staff sanitized their hands before and after entering patients' rooms. Once an electronic board was introduced, which displayed a positive message such as "Good job!" as response to an employee sanitizing her/his hands, the compliance rose to close to 90% within one month and was more likely to be repeated in the future. The researchers concluded that this response is hardwired in our brains. They argue that to reap rewards in life, like a promotion, we must take actions, and to avoid bad outcomes such as poison it is better to do nothing. This assumption is further supported by another function of our brain. We are more likely to believe and accept positive information than negative one. Both assumptions would explain why the positive, immediate feedback was more effective motivation than the threat of a probable illness.

Relating this to plastics recycling, it indicates that consumers will respond better to the promise of a cleaner environment when we reach our goals than to the threat of a polluted one if we do not.

Both the response to fines or a loss as well as to positive feedback are important aspects of our human behavior that should be considered when creating effective and actionable policies for plastics recycling and waste reduction.

5.1.4 Technology-Specific Instruments

Technology-specific instruments recognize that different recycling technologies have distinct needs and challenges. Mechanical recycling is well-established but limited by quality issues, whereas chemical recycling can process contaminated and mixed plastics but is still emerging [19].

Policy support may include:

- **Permitting fast-tracks:** Reducing regulatory barriers for chemical recycling pilot plants.

- **Public R&D funding:** Supporting the development of depolymerization, solvolysis, and pyrolysis technologies [20].

- **Technology-neutral design:** Establishing performance-based regulations that allow for innovation.

5.2 Policy Landscape across Countries and Regions

Policy approaches vary across regions depending on legislative traditions, public pressure, industrial capabilities, and political priorities. This chapter compares how different countries and regions influence plastics recycling through regulation, and how these policies shape the global process chain.

5.2.1 European Union

The EU has positioned itself as a global leader in circular economy and plastic waste legislation, using a combination of binding directives, economic instruments, and harmonized standards to drive systemic change. Its approach emphasizes product responsibility, sustainable design, and clear targets for recycling and waste prevention [21].

Framework Strategies:

■ **Circular Economy Action Plan (2020):** Introduces measures for sustainable product policy, waste prevention, and enhanced recycling [21].

■ **Packaging and Packaging Waste Regulation 2025/40 (PPWR):** Sets eco-design design requirements, minimum recycled content, and national recycling targets; replaces the old Packaging and Packaging Waste Directive 94/62/EC (PPWD) [3].

■ **Targeted Bans and Controls**

■ **Single-Use Plastics Directive (2019/904/EU):** Bans certain plastic items and mandates reduction targets for others [22].

■ **Market Based Instruments**

■ **EU Plastic Tax (2021):** Levies €800 per ton of non-recycled plastic packaging waste to incentivize recycling and fund the EU budget [9].

The EU's influence extends beyond its borders, as non-EU manufacturers must comply with EU rules to access the European market – driving international alignment through regulation.

5.2.2 United States

In the U.S., recycling policy is largely decentralized, with states taking the lead in regulatory development. This results in a patchwork of approaches, varying widely in ambition and enforcement.

State-Level Legislation

■ **EPR for packaging:** Enacted in Oregon, Maine, Colorado, California, and Minnesota, these laws shift financial responsibility to producers [23].

■ **Federal initiatives**

■ **National Recycling Strategy:** The U.S. Environmental Protection Agency (EPA) released a national framework in 2020 to improve recycling infrastructure, data collection, and stakeholder coordination nationwide [24].

■ **Voluntary commitments**

■ **U.S. Plastic Pact:** An industry-driven initiative to promote packaging recyclability and increase recycled content [25].

The U.S. model relies heavily on voluntary participation and state action, though momentum is growing for federal harmonization.

5.2.3 Asia-Pacific

Asia-Pacific countries face mounting pressure from both domestic waste volumes and global attention on plastic pollution. In response, many have implemented bold regulations to control waste imports and improve domestic recycling.

Trade and Import Restrictions

- **China's National Sword Policy:** Implemented in 2018, banned the import of most plastic waste, reshaping global recycling flows [26].

- **Producer Responsibility Models:**

- **Japan's Containers and Packaging Recycling Act:** A well-functioning EPR system with high public compliance and recovery rates [27].

- **India's Plastic Waste Management Rules:** Mandate EPR, progressive plastic bans, and recycling targets for producers [28].

Policy in the region varies in maturity, but the trend is toward stricter enforcement and domestic self-sufficiency in waste management.

5.3 Challenges and Gaps in Current Regulatory Frameworks

Despite notable progress, several challenges continue to hinder the effectiveness of current recycling policy frameworks [29]:

- **Fragmentation and inconsistency:** Varying rules across jurisdictions create confusion and hinder economies of scale [1].

- **Weak enforcement mechanisms:** Inadequate monitoring and penalties undermine compliance [30].

- **Technological mismatch:** Existing policies are often designed for mechanical recycling and may not account for chemical recycling technologies [19].

5.4 Role of International Agreements and Cooperation

Plastic pollution is a global issue that transcends borders. Multilateral cooperation is essential for managing waste flows and harmonizing standards.

- **Basel Convention:** Regulates international trade in plastic waste through consent-based protocols. (2021 plastic waste amendments) [31].

- **OECD guidelines:** Promote environmentally sound waste management and recycling practices among member countries [7].

- **G7 and G20 initiatives:** Support circular economy transitions and action against marine plastic pollution [32, 33].

Such agreements foster knowledge exchange and accountability, but greater alignment is still needed to reduce trade frictions and support global solutions.

5.5 Future Trends and Recommendations

The next decade will be critical in aligning environmental goals with economic and technological realities. As the complexity of plastic flows increases, future policy must become more integrated, science-based, and innovation-friendly.

Several policy trends are expected to shape the global landscape:

- **Digitalization:** Emerging technologies like digital product passports and blockchain will enable better material traceability and compliance [10].

- **Eco-modulation of EPR fees:** Differentiating fees by product recyclability or environmental performance will incentivize circular design [1].

- **Recycled content mandates:** Requirements for postconsumer recycled plastic (e. g., PET bottles) will support market development and stabilization [21].

- **Design for circularity:** Regulatory frameworks will increasingly link product design to end-of-life outcomes [34].

- **Technology-neutral innovation support:** Future policies may focus on performance-based outcomes (e. g., minimum recycling rates or GHG reduction thresholds) rather than mandating specific technologies [20].

- **International alignment:** Calls for standardized definitions, labeling, and reporting will intensify to streamline trade and compliance [2].

5.6 Recommendations for Policymakers

To support the ongoing transformation of plastic waste systems, the following actions are recommended by the authors:

- **Harmonize terminology and metrics** across jurisdictions to enable fair comparisons, inform better policymaking, and facilitate international trade [1].

- Implement **performance-based regulation** that encourages innovation in both mechanical and chemical recycling, rather than locking in specific technologies [19].

- Strengthen **enforcement mechanisms** to ensure that regulations achieve their intended environmental and economic outcomes [2].

- Expand **multilateral cooperation** to build capacity, address transboundary waste challenges and share best practices. Coordinated international collaboration is essential for aligning regulatory approaches, enabling technology transfer, and establishing shared standards for a global circular plastics economy. Existing international platforms should be further leveraged and developed to support this effort. [31].

- Promote **transparency and traceability** across the value chain through digital infrastructure, material tracking, and open-access data systems [10].

Ultimately, the success of future plastics policy will depend on the ability of governments, industries, and civil society to act collaboratively, globally, and with a shared commitment to long-term systems change.

References

[1] OECD. *Global Plastics Outlook: Policy Scenarios to 2060*. Paris, France. 2022.

[2] UN Environment Programme. *Single-use plastics: A roadmap for sustainability*. 2018.

[3] European Commission. *Packaging Waste*. Access Date: 21.04.2025. Available: *https://environment. ec.europa.eu/topics/waste-and-recycling/packaging-waste_en*.

[4] European Commission. *Turning the tide on single-use plastics*. Brussels. 2019.

[5] Kunststoffindustrie droht Rezyklat-Lücke. *K-Zeitung*. Access Date: 21.04.2025. Available: *https:// www.k-zeitung.de/kunststoffindustrie-droht-rezyklat-luecke*.

[6] Withana, S., Brink, P., Illes, A., Nanni, S., and Watkins, E. *Environmental tax reform in Europe: Opportunities for the future final report*. Institute for European Environmental Policy, Brussels. 2014.

[7] OECD. *Extended Producer Responsibility: Updated Guidance for Efficient Waste Management*. Paris. 2016.

[8] Kaza, S., Yao, L., Bhada-Tata, P., and Van Woerden, F. *What a waste 2.0: a global snapshot of solid waste management to 2050*. World Bank Publications. 2018.

[9] European Council. Council Decision (EU, Euratom) 2020/2053 of 14 December 2020 on the system of own resources of the European Union and repealing Decision 2014/335/EU, Euratom. Brussels, Belgium. 2020.

[10] European Commission. *Strategic EU Ecolabel. Work Plan 2020 - 2024*. Brussels. 2023.

[11] MacArthur, D. E., Waughray, D., and Stuchtey, M. R. *The New Plastics Economy: Rethinking the Future of Plastic*. Ellen MacArthur Foundation. 2016.

[12] Zero Waste Europe. *Zero Waste Europe Annual Report 2021*. 2021.

[13] Homonoff, T. A. Can small incentives have large effects? The impact of taxes versus bonuses on disposable bag use. *American Economic Journal: Economic Policy*. vol. 10. 2018. pp. 177–210.

[14] Sidhu, M., Mehrotra, K., and Hu, K. *Single-Use Items Reduction: Disposable Cups*. University of British Columbia. British Columbia, USA. 2018.

[15] Starbucks. Starbucks Rolls Out 5-cent Paper Cup Charge Across Germany. Access Date: 24.04.2025. Available: *https://stories.starbucks.com/emea/stories/2020/starbucks-rolls-out-5-cent-paper-cup-charge-across-germany/*.

[16] Lieberman, A. and Duke, K. Research: Why We're Incentivized by Discounts and Surcharges. Access Date: 24.04.2025. Available: *https://hbr.org/2020/02/research-why-were-incentivized-by-dis counts-and-surcharges*.

[17] Tversky, A. and Kahneman, D. Loss aversion in riskless choice: A reference-dependent model. In: *The Quarterly Journal of Economics*. vol. 106. 1991. pp. 1039–1061.

[18] Sharot, T. What Motivates Employees More: Rewards or Punishments? Access Date: 24.04.2025. Available: *https://hbr.org/2017/09/what-motivates-employees-more-rewards-or-punishments*.

[19] Closed Loop Partners. *Accelerating Circular Supply Chains for Plastics*. New York, USA. 2021.

[20] U.S. Department of Energy. *Strategy for Plastics Innovation*. Access Date: 21.04.2025. Available: *https://www.energy.gov/entity%3Anode/4394292/strategy-plastics-innovation*.

[21] European Commission. *A New Circular Economy Action Plan*. Brussels. 2020.

[22] European Parliament. Directive (EU) 2019/904 of the European parliament and of the council of 5 June 2019 on the reduction of the impact of certain plastic products on the environment. Brussels, Belgium. 2019.

[23] Product Stewardship Institute (PSI). EPR laws in the United States. Access Date: 21.04.2025. Available: *https://productstewardship.us/epr-laws-map/*.

[24] United States Environmental Protection Agency (EPA). National Recycling Strategy. Access Date: 21.04.2025. Available: *https://www.epa.gov/circulareconomy/national-recycling-strategy*.

[25] U.S. Plastics Pact. U.S. Plastics Pact Roadmap to 2025. Access Date: 21.04.2025. Available: *https:// usplasticspact.org/roadmap/*.

[26] Brooks, A. L., Wang, S., and Jambeck, J. R. The Chinese import ban and its impact on global plastic waste trade. *Science Advances*. vol. 4. 2018. eaat0131.

[27] Ministry of Environment Government of Japan. *Annual Report on the Environment in Japan 2022*. 2022.

[28] Ministry of Environment Forest and Climate Change. The Plastic Waste Management Rules. New Dehli, India. 2016.

[29] Ellen MacArthur Foundation. *Plastics and the circular economy – deep dive*. Access Date: 16.02.2025. Available: *https://www.ellenmacarthurfoundation.org/plastics-and-the-circular-economy-deep-dive*.

[30] UN Environment Programme. *Global Waste Management Outlook 2024*. 2024.

[31] Basel Convention. Amendments on Plastic Waste. Access Date: 21.04.2025. Available: *https://www. basel.int/default.aspx*.

[32] G7 Environment Ministers. *G7 Environment Ministers' Meeting*. Metz, France, 5–6 May, 2019. Communiqué. Metz, France. 2019.

[33] G20 Environment Ministers. *Osaka Blue Ocean Report: G20 Implementation Framework for Actions on Marine Plastic Litter*. Osaka, Japan. 2019.

[34] PlasticsEurope. *The Circular Economy for Plastics – A European Analysis*. Brussels, Belgium. 2024.

6

Plastic Waste around the World

As described in Chapter 1, there are three primary waste handling methods: landfilling, incineration with energy recovery (waste-to-energy – WTE), and recycling. Despite these methods being generally consistent across regions, the ratios of landfilling, WTE, and recycling differ significantly between countries and regions. As shown in Table 6.1, recycling rates in 2022 varied greatly between Asia, Europe, and North America. However, not only the recycling rates vary: Especially the amounts of plastic waste that remain uncollected vary significantly (from 0% in Japan to 28% in China) across these regions. In addition to political factors, described in Chapter 5, these differences are also driven by organizational and procedural reasons – both between regions and continents, and even within countries [1, 2].

Table 6.1 Overview of Plastic Waste Handling in Different Regions in 2022 [3, 4]

Region		Total Amount Plastic Waste [Mt / %]	Recycled Plastic Waste Mt / %]	WTE Plastic Waste [Mt / %]	Landfilled Plastic Waste [Mt / %]	Not Collected Plastic Waste [Mt / %]
Asia	China	66.2 / 100	15.5 / 23.4	22.3 / 33.7	9.80 / 14.8	18.6 / 28.1
	India	9.3 / 100	1.3 / 14.0	0.4 / 4.3	5.8/ 62.4	1.8 / 19.3
	Japan	9.2 / 100	2.8 / 30.4	5.2 / 56.5	1.2 / 31.1	0.0 / 0.0
Europe	EU27+3	33.3 / 100	8.7 / 26.1	16.0 / 48.0	7.4 / 22.2	1.1 / 3.3
North America	Canada	3.4 / 100	0.3 / 8.8	0.1 / 2.9	2.9 / 85.3	0.1 / 3.0
	USA	42.0 / 100	3.6 / 8.7	6.3 / 15.0	30.7 / 73.1	0.9 / 2.2

Therefore, in this chapter, the plastic waste handling processes of the different regions are analyzed in more detail. The chapter focuses on the regions and countries in Table 6.1, as these regions combine 60% of the global plastic waste production and provide a solid data base regarding plastic waste handling.

6.1 Europe

Europe has a highly developed plastics industry, which has remained relatively stable over the past decade. In 2022, Europe produced approximately 60 million tons of plastics, marking an increase of 5% since 2018. This rise in production reflects the growing demand for plastics. In 2022, plastic demand in Europe reached 58 million tons. In 2022, packaging accounts for the highest demand at 34.7%, followed by construction (23.5%) and automotive (8.8%) [3, 5].

With increasing plastics consumption, also the amount of waste increases. Packaging is again the dominating sector. However, its share in waste is 50% and thus significantly higher than its share in production. This difference arises due to the short service life of plastic packaging, especially single-use plastics (e. g., bottles, plastic bags, plastic wrap). In contrast, for example, plastics used in construction have a much longer service life of 30 to 40 years. This explains the discrepancy between the share of demand for plastic (23.5%) and plastic waste (9.0%) of the building and construction sector [6].

To manage the increasing plastic waste, all three presented waste management methods are used. However, their trends point in different directions. Whereas landfilling is constantly decreasing, recycling and WTE are constantly increasing, with WTE as currently dominating waste management method. In 2016, for the first time, more plastic was recycled than landfilled in Europe. In 2022, 32.2 million tons of plastic waste were collected through official schemes (out of 33.3 Mt plastic waste in total), from which 49.6% were burned to recover energy, 26.9% were recycled, and 23.5% were landfilled. Figure 6.1 shows the development of waste treatment in Europe between 2006 and 2022 [3].

However, this development is not the same in every country. Due to landfill bans in Switzerland, Austria, Netherlands, Germany, Sweden, Luxembourg, Denmark, Belgium, Norway, and Finland, which started partially in 1996, WTE and landfilling are dominating in these countries. There, the landfilled plastic is at nearly zero percent. However, on the contrary, Croatia, Bulgaria, Cyprus, Greece, and Malta have landfill ratios for plastic waste of over 60% [2]. Figure 6.2 shows the different plastic waste handling ratios of the EU27+3 region.

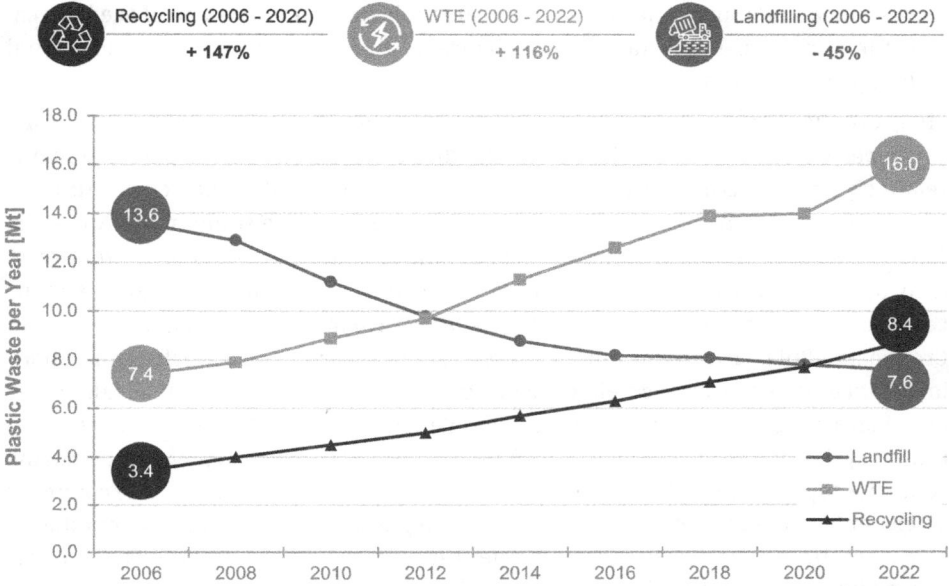

Figure 6.1 Plastic waste treatment in the EU27+3 countries from 2006 to 2022 [3]

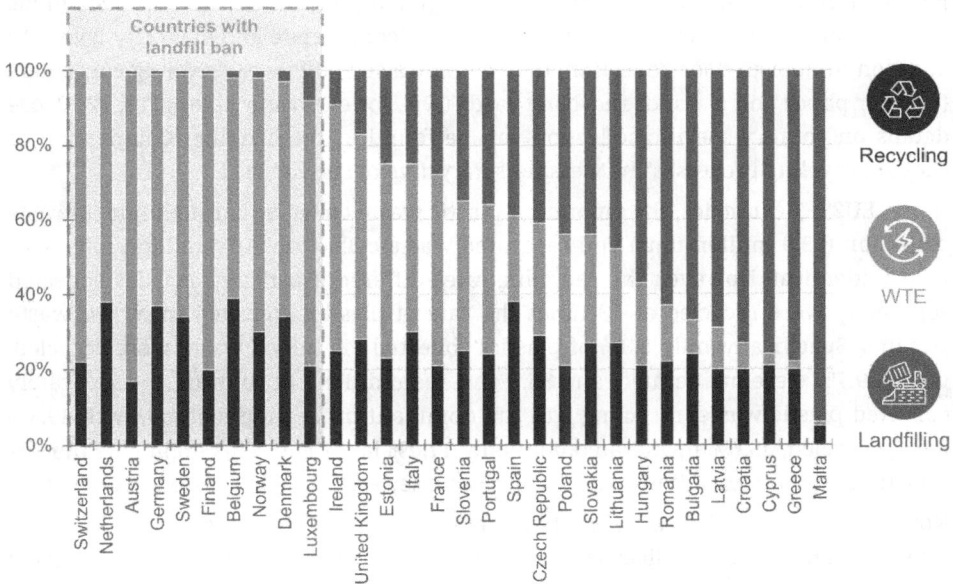

Figure 6.2 Waste handling in the different countries in 2022 (sorted by increasing landfilling share from left to right) [3]

As described in Chapter 5, legal and political regulations comprise one of the key reasons for disparities in plastic recycling between countries. However, two additional factors play an important role.

First, Croatia, Greece, Bulgaria, and Cyprus generate the lowest amount of *plastic waste per capita* in the EU [7]. This means that the total amount of plastic waste is relatively small, making it economically unfeasible to establish dedicated plastic recycling facilities. As analyzed in Chapter 4 and shown by a case study in the Appendix (Chapter 8), a minimum threshold of plastic waste is necessary to ensure that recycling operations are at least economically viable, if not profitable [8]. Second, all four countries have a *gross domestic product (GDP) per capita* below the EU average, with Croatia and Bulgaria ranking among the lowest [9]. According to a World Bank report, lower income levels are correlated with lower rates of recyclable materials in the waste stream, compounding the first issue. Furthermore, waste collection rates and thus recycling rates vary widely based on income. High- and upper-middle-income countries typically provide universal waste collection, whereas low-income countries collect only about 50% of waste in cities and roughly 25% in rural areas. In middle-income countries, rural waste collection rates range from 33% to 45% [10]. These differences directly impact the plastic recycling rate.

Different types of *waste collecting systems* exist across countries. The design of these systems – whether complete separation of recyclable materials, partial separation, or joint collection – depends on national and regional policies. Germany, for example, has different waste collection approaches at the federal state and even city levels. In addition to door-to-door collection, they employ also stationary recycling containers (e. g., for paper and glass) or recycling yards (e. g., for electronic scrap) [11, 12]. More details on country-specific collection can be found in the chapter "Comparison of Global Recycling Processes" in Niessner's *Recycling of Plastics* [11].

In the EU27+3 countries, the amount of plastic waste collected through mixed waste collection (15.9 million tons) and separated waste collection (16.4 million tons) was nearly identical. However, the recycling rates differed drastically. Plastics collected separately were recycled at 13 times the rate of plastics collected in mixed waste streams. Specifically, only 3.8% of plastics collected via mixed waste were recycled, while 59.7% were incinerated, and 36.5% were landfilled. In contrast, for separately collected plastic waste, recycling was the dominant processing method, with 49.4% being recycled, 39.6% incinerated, and only 11.0% landfilled. This highlights the significant environmental impact of proper waste separation, as discussed in Chapter 4.

Not all recycled plastic is converted into new *recyclates*. Some of the processed recycled material is used as pellets, while a portion is exported outside the EU, re-entering the global economy. In 2022, the EU exported 1.1 million tons of plastic waste – representing a 67% decrease compared to 2014, when exports peaked at 3.24 million tons [3, 13]. The final destinations of exported plastics are further analyzed in Section 6.3 of this book.

6.2 North America

The increase in plastic production and the replacement of other materials in many industries is also leading to an increasing amount of plastic waste in North America, respectively NAFTA. The increased consumption results in an increased plastic waste amount. However, here lies a difference compared to Europe. The treatment of plastic waste differs significantly, as the following sections show for the USA (Section 6.2.1) and Canada (Section 6.2.2).

6.2.1 USA

The total amount of plastic waste in the United States increased by a factor of 80 between 1960 and 2017. In that time frame, the share of plastic waste in municipal waste also grew in parallel, demonstrating the increasing importance of plastic waste treatment in the United States. However, there is a clear difference compared to Europe in terms of plastic waste treatment. As shown in Section 6.1, especially in Figure 6.1 , landfilling of plastic waste in Europe is also decreasing and accounted for the smallest share of plastic waste treatment at 24.7% in 2018. Recycling accounted for 32.3% in Europe in the same year, and incineration with energy recovery 43.0%. Until 2015, the recycling rate in the USA increased gradually. However, since 2015, this development has stagnated and is possibly even regressive. In 2022, landfilling dominated waste management of plastic waste at 75.0%. Incineration with energy recovery accounted for 16.3% and recycling for only 8.7%. The evolution of the distribution of the three main waste management measures from 1960 to 2022 is shown in Figure 6.3 [14, 15].

Unlike the European recycling process, single-stream recycling is the standard approach in the U.S. In this system, recyclable materials such as paper, glass, plastics, and metal are primarily collected through door-to-door collection and then transported to material recovery facilities (MRFs). An MRF is a processing facility where waste is sorted and prepared for further processing. Plastics, for example, are separated from other materials and sent to specialized plastic processing facilities. Single-stream recycling was first introduced in Phoenix and Arizona, in 1989, with the first large-scale MRF beginning operations in 1993 [16]. Over time, many U.S. states and cities transitioned from curbside waste collection to single-stream recycling. In 2018, around 80% of U.S. communities had adopted this system. Across the country, 650 MRFs process up to 100 kt plastic waste daily [1].

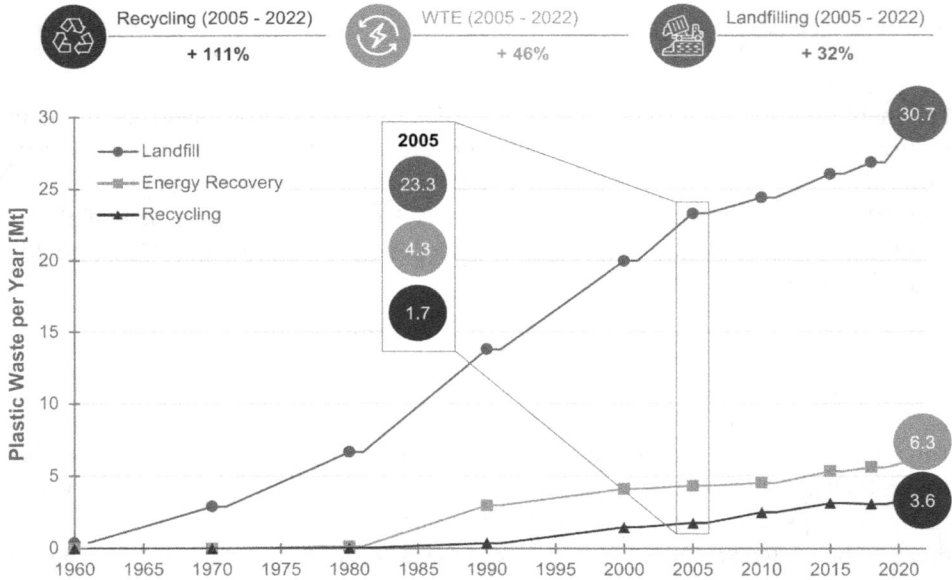

Figure 6.3 Plastic waste treatment in the USA from 1960 to 2022 [14, 15]

One of the major challenges of single-stream recycling is *contamination*, with approximately 25% of plastic materials being affected (as described in Chapter 4 and shown for the EU in Section 6.1). Contaminated materials cannot be recycled and are instead either incinerated or sent to landfills. This contamination reduces the economic viability of recycling and, in many cases, makes it unprofitable for recycling companies. The financial impact varies by state, but in many regions, relatively low landfill fees for mixed plastics further discourage recycling efforts [1]. The trends shown in Figure 6.3 reflect this situation.

Another key issue is the heavy emphasis on *PET recycling* in the USA. PET is prioritized because it is relatively easy to recycle into new products such as plastic bottles and packaging. Aside from industrial materials (mainly non-PET resins), PET is the only plastic resin with a double-digit recycling rate. Given that only PET and the "other" resins are considered economically viable for recycling, only about 25% of material end up being recycled. These three factors – high contamination rates, inexpensive landfill disposal, and a low overall recycling rate – reinforce one another, leading to stagnation or even decline in plastic recycling.

To improve recycling rates, 10 states have implemented beverage container deposit laws (similar to Europe, particularly in Germany). However, the number of states with such laws has remained unchanged since 2002 [17].

Similar to the EU27+3 countries, the U.S. does not reuse all of its recyclates. In fact, a significant portion of recycled materials has historically been exported. The peak of these exports occurred in 2015, with approximately 2 million tons of plastic waste (equivalent to 63% of all recycled plastics) being sent abroad. A notable shift in export patterns took place between 2017 and 2018, as illustrated in Figure 6.4. This change was driven primarily by China's import ban on plastic waste, which is discussed in more detail in Section 6.3. As a result, U.S. plastic waste exports to China dropped by almost 90%. Although shipments to alternative destinations such as Malaysia, Thailand, and Vietnam increased, the total volume of exported plastic waste in the first half of 2018 was still about one-third lower than in the same period of 2017. This shift created a major challenge for the U.S. recycling industry, leaving approximately 600,000 tons of plastic waste without an export destination in 2018. The resulting surplus drove down the market value of recycled plastics, further reducing the economic feasibility of recycling in the U.S. This downward trend has persisted, with export volumes continuing to decline. By 2023, only 0.4 million tons were exported, which is only 20% of the volume exported in 2015. Given these challenges, the U.S. must develop new solutions for plastic waste management or work toward reducing plastic waste altogether. Other countries, particularly in Europe, have already implemented measures that could serve as useful models for addressing this issue in the United States [18, 19].

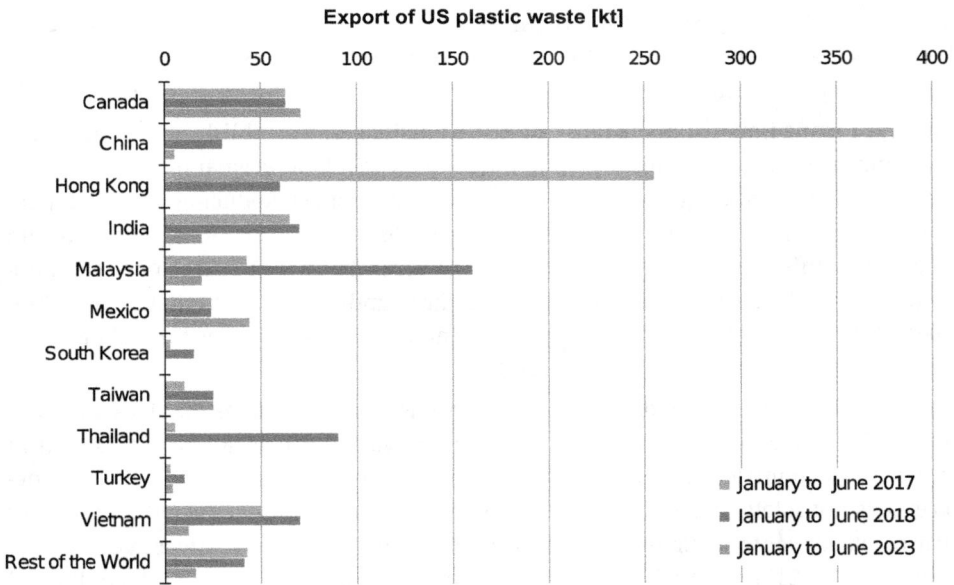

Figure 6.4 Distribution of U.S. plastic waste exports by country [18, 20]

6.2.2 Canada

In 2016, Canada generated approximately 3.24 Mt of plastic waste, equivalent to 118 kg plastic waste per capita. This is slightly higher than the amount per capita in the USA (105 kg) [1]. Three sectors accounted for nearly 70 percent of plastic consumption: packaging (33%), construction (26%), and automotive (10) [21]. A similar trend to Europe can be observed in waste generation: packaging accounted for 47% of all discarded plastic waste, significant exceeding its share of plastics use in the same year [21].

Of the plastic waste generated and collected in 2016, the vast majority ended up in landfills (87%), while 4% was converted to energy through WTE-facilities. Only 9% was recycled, with most undergoing mechanical recycling (8% of total collected plastic waste) and a small fraction processed chemically (1% of total collected plastic waste) [21].

Several process-related and organizational challenges contribute to Canada's relatively low recycling rate and high reliance on landfilling. One key issue lies in the *collection system*: Improper sorting at both the consumer and collection level – partially due to the increasing use of single-stream collection systems – leads to contamination of collected plastics. This requires additional sorting and quality control at MRFs, as well as specialized technologies to remove impurities. In many cases, contaminated plastic cannot be reused as recyclates.

Beyond sorting issues, Canada faces logistical challenges in establishing a comprehensive and cost-effective collection system due to its vast land area and widely distributed waste generation (e. g., end-of-life agriculture plastics). Reducing landfill dependence would require numerous small-scale MRFs or WTE facilities, which would require significant investment (cf. Chapter 4). Currently, only 11 MRFs operate nationwide, forcing large volumes of plastic waste into landfills. In addition, Canada's limited *WTE infrastructure* further restricts alternative disposal options for hard-to-recycle plastics, leading to even more landfilling [21].

Like the EU and USA, Canada also exports the waste and recyclates that lack a domestic market. Canada has experienced a decline in waste exports, along with a shift in destination countries, mirroring trends in the USA (see Figure 6.4). However, the decrease is not solely due to China's import ban, but also the *Basel Convention's plastics waste amendments* (cf. Chapter 5). These amendments, which took effect in Canada on January 1, 2021, regulate international shipments of plastic waste that have limited recycling options or are difficult to recycle. These include contaminated plastics (e. g., residues remaining in containers or mixed with other household wastes), halogenated plastics, and multi-polymer plastic wastes. These changes will likely impact all stakeholders in the supply chain, including collectors and processors, who may need to adjust their collection, sorting and processing methods to meet new market requirements [22].

6.3 Asia

The region "Asia and Oceania" produces the highest amount of plastic waste, as Figure 1.1 shows. However, this number must be put into perspective with its total population, which is nearly five billion. This means, that the plastic waste per capita in this region is 43.9 kg and thus less than half of the amount per person in the USA or Canada. This number varies a lot between the countries, which depends on different factors (e. g., income level, regulations). Therefore, the following subchapters analyze China (Section 6.3.1), India (Section 6.3.2), and Japan (Section 6.3.3) in detail. Section 6.3.4 points out highlights of additional Asian countries.

For more than 25 years, Asian countries have played an important role in the global plastic waste management value chain, as Western countries export to economically disadvantaged Asian countries. However, many of these countries lack the infrastructure and capacity to properly manage such waste [23]. Section 6.1 and Section 6.2 have already analyzed these value chains from the perspective of exporting nations. The following subchapter will now examine the impact on importing countries, with a particular focus on the shift in plastic imports following China's import ban in 2018.

It is important to note that data availability and reliability for Asian countries are not as strong as for Europe or North America. As a result, this analysis is, at times, more qualitative than quantitative.

6.3.1 China

As consequence of China's fast economic growth, the country overtaken all other countries and regions in 2013 to become the world's largest producer of plastic products. Its domestic consumption of plastics has skyrocketed as well [8]. Since then, China is both the largest producer and consumer of plastics and plastic products. In 2018, China produced 83.6 Mt of plastic waste and consumed 67.6 Mt [1].

The *plastic consumption* by sector is not as well documented as in Europe and North America. The same applies to data on recycling, WTE and landfilling. As shown in Table 6.1, a study of conversion estimates, from 40.1 Mt of collected plastics (68.8% of total produced waste) in 2018, 31% were recycled, 48% were incinerated with energy recovery and 21% were landfilled. However, more than 30% of the plastic waste generated was not managed at all.

In-country *collection and recycling* in China is much more difficult than in Europe and the U.S. In many areas, waste management infrastructure is still underdeveloped. There is no nationwide formal recycling system in place – neither for plastics nor for other recyclables. The largest share of plastic waste is still collected by informal waste pickers, who collect plastic bottles and earn a small amount of money in return. As a result, many cities rely on private waste management companies that often lack adequate recycling strategies [1].

However, China has been steadily implementing recycling systems. On July 1, 2019, Shanghai became the first city to launch a mandatory waste separation at source. Whereas previously all MSW was disposed of in a single bin, Shanghai has now added four separate bins: one for dry residual waste, one for wet residual waste, one for hazardous materials, and one for recyclables. These bins are emptied daily. In addition, according to media reports, around 30,000 volunteers have been mobilized to educate Shanghai residents on waste separation. The first positive effects were quickly observed, particularly in reducing street litter. One year after implementation, compared to 2018, the average amount of recyclables collected daily in 2019 increased by 431.8%. The amount of kitchen waste and hazardous waste collected also grew by 88.8% and 504.1%, respectively, while residual waste decreased by 31%. Although specific data on plastic waste is not yet available, a positive trend is evident. The success of Shanghai's complex waste segregation regulations has influenced other Chinese cities, including Guangzhou, Kunming, Xi'An and Beijing, to adopt mandatory garbage sorting for households and businesses [24]. This shift das also impacted recycling processes and waste handling.

Analyzing the Chinese plastic recycling chain, *recyclates* play an important role. Unlike North America or Europe, China's economic growth has historically been linked to the import of plastic waste. Due to low-quality domestic postconsumer plastic waste sources and the high demand for plastics, China became the world's largest importer of plastic waste beginning in the late 1980s. Between 1988 and 2016, China (including Hong Kong) imported 170.5 Mt of plastic waste – accounting for 72% of the total global plastic waste imports. Although China was also a major exporter of plastic waste, with a cumulative weight of 56.1 Mt and a share of 26.1%, its net imports were significantly higher. Table 6.2 summarizes plastic waste imports and exports for selected countries between 1988 and 2016 [25]. Among these countries, China is the only one with more imports than exports. In contrast, Germany, Japan, and the USA had the lowest net imports, making them the "net-export champions" of plastic waste.

Table 6.2 Plastic Waste Imports and Exports by Country between 1988 and 2016 [25]

Country	Import		Export		Net Import [Mt]
	Cumulated Amount [Mt]	Global Share [%]	Cumulated Amount [Mt]	Global Share [%]	
Canada	3.83	1.62	3.89	1.81	−0.06
China	170.50	72.4	56.1	26.10	114.4
Germany	5.36	2.27	17.6	8.22	−12.24
Japan	negl.	negl.	22.2	10.30	−22.2
USA	8.49	3.60	26.7	12.40	−18.21

Since 2013, however, China's import quota for postconsumer plastic waste has declined despite rising demand. This trend began with Operation Green Fence in 2010, which aimed to reduce low-quality plastic waste imports by requiring pre-sorted plastics by type. As a result, China imported 2 million tons of postconsumer plastic waste from January to April 2016, an 11% decrease from the same period in 2015 [25, 26, 27].

By 2027, China recognized the potential to meet its postconsumer plastic demand domestically by 2030 and tightened regulations due to increasing pollution from low-quality plastic imports. A stricter, permanent ban on non-industrial plastic imports was introduced. In the first year of the ban, the impact was clear: As shown in Figure 6.4 the U.S. plastics exports to China dropped from 380,000 tons (January to June 2017) to 35,000 tons (January to June 2018), while UK exports fell from around 90,000 tons (January to April 2017) to around 1,500 tons (January to April 2018). Some plastics waste was redirected to other Asian countries, such as Malaysia, Taiwan, Vietnam, and Indonesia) [18, 25, 28, 29]. Table 6.3 summarizes changes in EU exports between 2015 (pre-ban) and 2018 (post-ban) for select Asian countries.

Despite the ban, Japan, Germany, and the USA remained the top plastic waste exporters in 2019, while China was no longer among the top five net importers. Instead, Turkey, Vietnam, and Malaysia had the highest net imports that year [30].

Table 6.3 Import Changes after China Ban from EU [25]

Country	EU Imports 2015		EU Imports 2018		Changes [%]
	Cumulated Amount [t]	EU Share [%]	Cumulated Amount [t]	EU Share [%]	
China	1,658,970	54	64,662	3	−96
Hong Kong	775,557	25	211,530	11	−73
India	139,628	5	158,250	8	+13
Indonesia	32,640	1	190,933	10	+485
Malaysia	137,876	4	404,123	21	+193
Taiwan	31,548	1	99,071	5	+214
Thailand	15,414	1	39,676	2	+157
Turkey	19,377	1	270,339	14	+1,295
Vietnam	88,760	3	187,378	10	+111
Other	177,742	6	301,135	16	+69
Total	**3,077,512**		**1,927,097**		**−37**

In 2022, the plastic waste import to China further decreased to an insignificant 0.07 million tons. This marks a decrease of 98% imported plastic waste within 10 years. China thereby developed from the top import country in 2017 (5 million tons, 37% global import volume) to a nearly negligible player, covering 1.1% of the global plastic waste imports. China set an example for many other countries: Since 2017, global plastic waste import and export volumes have declined from 12.4 to 6.3 million tons (nearly a 48% reduction) [31]. As a result, countries must now focus on improving their own waste management systems and closing their loops to advance a circular economy (cf. Chapter 1).

6.3.2 India

Despite India's population being nearly as large as China's, its *plastic consumption* is only a fifth of China's. However, due to overall development and especially the growing construction sector, demand exceeds domestic production. This is reflected in India's plastic production growth from 8.3 million tons in 2010 to 22 million tons in 2020. The demand is primarily for short-lifetime plastics (e. g., packaging, bottles, plastic bags) [32].

The lower plastic consumption is also reflected in a smaller amount of plastic waste, which totaled 9.3 million tons in 2022. However, plastic waste management data quality is lacking. While India reports a national waste collection coverage of around 95%, a study by researchers at the University of Leeds found that official statistics exclude rural areas, open burning of uncollected waste, and waste recycled by the informal sector. According to the study, only 81% of waste is actually collected [33, 34].

Of the 9.3 Mt of plastic waste generated per year, 81% is collected, meaning a fifth is dumped in oceans or unmanaged landfills. Additionally, 5.8 million tons end up in open landfills and are burned without energy recovery, causing severe environmental pollution. India has yet to implement systematic collection and recycling systems. However, due to low labor costs and an existing domestic markets for recycled plastics, its informal collection and recycling industry is relatively large compared to other emerging countries. This informal recycling focusses on PET and HDPE bottles and other rigid plastics, making India the emerging country with the highest recycling rate for these materials [1].

India's *recyclate* usage ratio is relatively high due to strong plastics material demand. Furthermore, in 2029, India was the fifth-largest net importer of plastics waste, with imports exceeding exports by 24 kt [30].

6.3.3 Japan

In 2017, Japan consumed around 10 Mt of waste, with 41% used for packaging, the plastic consumption in Japan was similar to Western countries [1].

That same year, Japan generated approximately 9 Mt of plastic waste, all of which was formally collected. The reason for this high collection rate is the multi-stream separation at the household level, where end users sort plastic, paper, PET bottles, aluminum, and glass for separate collection and recycling [1].

Japan employs all three major plastic waste handling treatment methods. Of the collected waste, 14% was landfilled, 56% was incinerated with energy recovery (WTE), and 30% was recycled. Some sources report a recycling rate of over 80%, but these figures include WTE as "recycling". Compared to other countries, Japan has a relatively high rate of chemical recycling. Of the 3 Mt recycled plastic waste, 0.4 Mt (13%) were chemically recycled, while 2.6 Mt (87%) were mechanically recycled into recyclates [1].

A significant share of these recyclates is exported. Japan previously exported large amounts to China, leading to a decline in total exports from 1.5 Mt in 2015 to 0.898 Mt in 2029 [35]. By 2022, this figure had further decreased to 0.56 million tons. Despite this decrease, Japan remains the third-largest plastic waste exporter globally and ranks fourth in net exports, with 0.06 million tons more exported than imported [31].

6.3.4 Other Asian Countries

As Table 6.3 shows, the import ban in China affects the plastic imports of other countries. Shifts in European waste have also been observed in other regions.

Since 2018, **Malaysia** has become a major import hub of plastic waste, importing 0.7 Mt of plastic waste. Compared to the plastic waste Malaysia produced domestically (1.8 Mt), this import amount was relatively high (40%). The recycled materials, including PET, LDPE, HDPE, and rigid plastics, are processed in various recycling facilities, leading to a recycling rate of 18%. However, Malaysia does not operate any WTE plant, meaning the remainder of the waste is landfilled, often unmanaged. As a results, 39% of the plastics are mismanaged and often end up in the oceans. Consequently, exports to Malaysia must be carefully monitored [1].

Indonesia has also reported a significant increase in plastic waste imports (> 400%). Despite this, the country faces considerable challenges in waste collection and management. In 2018, less than half of the waste was collected (46%), with 30% ending up in landfills. While 16% was recycled, not all of it was mechanically recycled, and some was used as alternative fuel [1].

These examples highlight the difficulties faced by Asian countries importing plastic waste from high-wage countries. Without proper waste management infrastructure, the import of plastic waste will only exacerbate existing challenges.

6.4 Summary and Conclusion

Plastic waste management remains a critical global challenge, marked by significant regional differences in collection, recycling, and disposal practices. While some regions have advanced waste treatment systems, others struggle with ineffective management, leading to high levels of landfilling and environmental pollution.

Europe has made substantial progress in reducing landfilling through increasing recycling and waste-to-energy (WTE) incineration. However, recycling rates vary between countries, influenced by economic factors and waste collection systems. The EU's regulatory framework has been instrumental in improving plastic waste management.

North America presents a stark contrast in plastic waste handling. Whereas Canada and the United States generate significant plastic waste, both countries rely heavily on landfilling. Recycling rates are low due to contamination issues, lack of infrastructure, and economic inefficiencies in the recycling market. The U.S. also faces challenges from its single-stream recycling system, which reduces the quality of recyclables.

Asia accounts for the largest share of global plastic waste production but varies greatly in waste management efficiency. Japan has a high collection and recycling rate, whereas China has drastically shifted its policies, banning plastic waste imports to focus on domestic waste management. India struggles with waste collection in rural areas, and many other developing Asian nations face high levels of mismanaged plastic waste due to insufficient infrastructure.

One of the major global shifts in plastic waste management occurred with China's 2018 import ban, disrupting global recycling markets and forcing many Western countries to seek alternative solutions. However, this has also led to waste being redirected to other Asian countries with less-developed waste management systems, further exacerbating environmental issues.

The global plastic waste crisis demands immediate, coordinated action. While high-income countries have invested in advanced recycling and WTE technologies, developing nations continue to struggle with inefficient waste collection and treatment. The shift in plastic waste exports has exposed the limitations of relying on foreign waste processing, underscoring the need for stronger domestic waste management policies.

To effectively tackle plastic pollution, the following key measures must be prioritized:

- Strengthening circular economy approaches – Increasing the use of recycled plastics in manufacturing and reducing single-use plastics can help minimize waste generation.

- Improving waste collection and sorting systems – Proper waste segregation and investment in infrastructure can enhance recycling efficiency and reduce landfilling.

- Policy and regulatory reforms – Countries must implement stricter regulations on plastic production, waste exports, and recycling standards to ensure sustainable waste management.

- Investment in innovative technologies – Advancements in chemical recycling, biodegradable plastics, and alternative materials can reduce plastic dependency.

- Global cooperation – As plastic waste is a transboundary issue, international collaboration is essential to develop effective waste management solutions and prevent plastic pollution in oceans and ecosystems.

While progress has been made, much remains to be done to establish a truly sustainable global plastic waste management system. Achieving this goal requires a combination of technological innovation, policy enforcement, and behavioral changes at both the industrial and consumer levels. This will be further investigated in Chapter 7 of this book.

References

[1]   Conversio Market & Strategy GmbH. *Global Plastic Flow 2018*. Mainaschaff, Germany. 2020.

[2]   PlasticsEurope. *Plastics – the Facts 2020: An analysis of European plastics production, demand and waste data*. Brussels, Belgium. 2021.

[3]   PlasticsEurope. *The Circular Economy for Plastics – A European Analysis*. Brussels, Belgium. 2024.

[4]   Statista. Plastikmüll. Access Date: 16.03.2025. Available: *https://de.statista.com/themen/4645/plastikmuell/#topicOverview*.

[5]   PlasticsEurope. *Plastics – the fast Facts 2024*. Brussels, Belgium. 2024.

[6]   European Court of Auditors. *EU Action to Tackle the Issue of Plastic Waste*. Luxembourg. 2020.

[7]   Statista Research Department. Generation of plastic packaging waste per capita in the EU 2018 by country. Access Date: 02.08.2021. Available: *https://www.statista.com/statistics/972604/plastic-packaging-waste-generated-per-capita-countries-eu/*.

[8]   Rudolph, N., Kiesel, R., and Aumnate, C. *Understanding Plastics Recycling: Economic, Ecological, and Technical Aspects of Plastic Waste Handling*, 2nd ed., Hanser: Munich. 2020.

[9]   Statista. Bruttoinlandsprodukt (BIP) pro Kopf in den EU-Ländern 2020. Access Date: 02.08.2021. Available: *https://de.statista.com/statistik/daten/studie/188766/umfrage/bruttoinlandsprodukt-bip-pro-kopf-in-den-eu-laendern/*.

[10] Kaza, S., Yao, L., Bhada-Tata, P., and Van Woerden, F. *What a waste 2.0: a global snapshot of solid waste management to 2050*. World Bank Publications. 2018.

[11] Kiesel, R. Comparison of Global Recycling Processes. In: *Recycling of Plastics*. Niessner, N., Ed. Hanser: Munich. 2022.

[12] Seyring, N., Dollhofer, M., Weißenbacher, J., Bakas, I., and McKinnon, D. Assessment of collection schemes for packaging and other recyclable waste in European Union-28 Member States and capital cities. *Waste Management & Research*. vol. 34. 2016. pp. 947–956.

[13] Tiseo, I. Annual plastic waste exports from the EU-27 2005–2020. Access Date: 02.08.2021. Available: *https://www.statista.com/statistics/1235915/plastic-waste-exports-european-union/*.

[14] United States Environmental Protection Agency (EPA). *Advancing Sustainable Materials Management: 2018 Tables and Figures Assessing Trends in Material Generation, Recycling,Composting, Combustion with Energy Recovery and Landfilling in the United States*. 2020.

[15] United States Environmental Protection Agency (EPA). *Advancing Sustainable Materials Management: 2018 Fact Sheet - Assessing Trends in Material Generation, Recycling, Composting, Combustion with Energy Recovery and Landfilling in the United States*. 2020.

[16] Guttentag, R. Processing recyclables: What's my line? *World Wastes*. vol. 37. 1994. pp. 28–35.

[17] National Conference of State Legislatures (NCSL). State Beverage Container Deposit Laws. Access Date. Available: *https://www.ncsl.org/research/environment-and-natural-resources/state-beverage-container-laws.aspx*.

[18] Clarke, J. S. and Howard, E. US Plastic Waste Exports to Developing Countries, Causing Environmental Problems at Home and Abroad. Access Date: 04.08.2021. Available: *https://unearthed.greenpeace.org/2018/10/05/plastic-waste-china-ban-united-states-america/*.

[19] Dell, J. 157,000 Shipping Containers of U.S. Plastic Waste Exported to Countries with Poor Waste Management in 2018. Access Date: 04.08.2021. Available: *https://www.plasticpollutioncoalition.org/blog/2019/3/6/157000-shipping-containers-of-us-plastic-waste-exported-to-countries-with-poor-waste-management-in-2018*.

[20] Basel Action Network (BAN). US Export Data - 2023 Summary. Access Date: 16.03.2025. Available: *https://www.ban.org/plastic-waste-project-hub/trade-data/usa-export-data-annual-summary#:~:text=U.S.%20PVC%20Plastic%20waste%20exports,million%20kg%2Fyr%20in%202023*.

[21] Deloitte. Economic Study of the Canadian Plastic Industry, Markets and Waste: summary report. Environment and Climate Change Canada. Ottawa, Canada. 2019.

[22] Government of Canada. Basel Convention plastic waste amendments: Impacts on Canadian recycling industry. Access Date: 03.08.2021. Available: *https://www.canada.ca/en/environment*

-climate-change/services/managing-reducing-waste/permit-hazardous-wastes-recyclables/basel-con vention-plastic-waste-amendments-impacts-industry.html.

[23] Parker, L. and Elliot, K. Plastic Recycling Is Broken. Here's How to Fix It. Access Date: 04.08.2021. Available: *https://www.nationalgeographic.com/science/graphics/china-plastic-recycling-ban-soluti ons-science-environment.*

[24] Green Initiatives. One year of waste segregation in Shanghai: Success or Failure? Access Date: 04.08.2021. Available: *https://greeninitiatives.cn/one-year-of-waste-segregation-in-shanghai-success- or-failure/.*

[25] Brooks, A. L., Wang, S., and Jambeck, J. R. The Chinese import ban and its impact on global plastic waste trade. *Science Advances.* vol. 4. 2018. eaat0131.

[26] Koty, A. C. Trash or Treasure? – Prospects for China's Recycling Industry. *China Briefing.* 2016. Access Date 27.05.2025. Available: *https://www.china-briefing.com/news/trash-or-treasure-prospects- for-chinas-recycling-industry/.*

[27] Waste and Resource Action Programme (WRAP). Plastics Market Situation Report. In: 2016. pp. 1–29.

[28] Deutsche Welle. After China's import ban, where to with the world's waste? Access Date: 04.08.2021. Available: *https://www.dw.com/en/after-chinas-import-ban-where-to-with-the-worlds-waste/a-48213871.*

[29] Financial Times. Plastic waste export tide turns to south-east Asia after China ban. Access Date: 04.08.2021. Available: *https://www.ft.com/content/94ee72d0-6f26-11e8-852d-d8b934ff5ffa.*

[30] Buchholz, K. *Which Countries Export & Import Plastic Waste?* Access Date: 04.08.2021. Available: *https://www.statista.com/chart/18229/biggest-exporters-of-plastic-waste-and-scrap/.*

[31] Park, B. C., Brown, A., Laubinger, F., and Börkey, P. Monitoring trade in plastic waste and scrap. *OECD Environment Working Papers.* 2024.

[32] Padgelwar, S., Nandan, A., and Mishra, A. K. Plastic waste management and current scenario in India: a review. *International Journal of Environmental Analytical Chemistry.* 2019. pp. 1–13.

[33] Velis, C. A. and Cook, E. Mismanagement of plastic waste through open burning with emphasis on the global south: a systematic review of risks to occupational and public health. *Environmental Science & Technology.* vol. 55. 2021. pp. 7186–7207.

[34] Prakash, P. India, not China, is world's largest plastic emitter: study Access Date: 16.03.2025. Available: *https://www.thehindu.com/sci-tech/energy-and-environment/india-is-the-worlds-largest- plastic-polluter-according-to-new-study/article68621895.ece.*

[35] Klein, C. Plastic waste export volume Japan 2016–2020. Access Date. Available: *https://www.statista. com/statistics/1193746/japan-plastic-waste-export-volume/.*

7

From Recycling to a True Circular Economy

Plastic recycling has entered a decisive phase. While we must continue to address the challenges posed by legacy products already in circulation, the real focus must shift toward designing future products for true circularity. Achieving a genuinely sustainable plastics economy requires more than improving recycling rates – it demands a systemic transformation in how plastics are designed, manufactured, used, and recovered.

Circularity must become both affordable and sustainable: it must offer predictability for recyclers, manageable documentation requirements, and economic viability at scale.

Governments must take an active role in providing the infrastructure necessary for documentation and material tracking, for example by supporting the implementation and harmonization of Digital Product Passports (DPPs). These digital systems will be crucial for ensuring transparency, improving material flows, and protecting intellectual property and sensitive corporate knowledge while still providing the critical recyclability information needed at end-of-life.

Ultimately, true circularity must prioritize quality, traceability, and global collaboration over mere appearance or greenwashing. Not all applications will be perfectly recyclable – and where recycling is not the most sustainable solution (e. g., with fiber-reinforced composites), alternative strategies must be pragmatically accepted.

7.1 Our Vision: Honest, Practical, Collaborative

A functioning circular economy begins with a realistic view of materials and processes. That means putting honesty over greenwashing. The goal is not to promote the supposedly most sustainable solution, but to choose the technically and ecologically best option for each application – based on reliable data about environmental performance.

At the same time, we need a future-focused design mindset: Products must be conceived from the outset to be recyclable, repairable, or reusable. Only then can we create material cycles that are viable in the long term. But vision alone is not enough – implementation must also be economically feasible and predictable. Recyclates will only gain widespread adoption if they are affordable for recyclers and processors and consistent in their performance.

Another key element is smart documentation infrastructure. Authorities and industry should work together to establish standardized systems that make relevant material data available throughout the product life cycle – especially at end of life. At the same time, manufacturers' intellectual property must be protected.

Quality and traceability are essential. True sustainability is not measured by the volume recycled, but by how high-quality and transparent the material flows are. And finally, it is clear that a truly circular plastics economy can only be achieved through global collaboration. Global challenges demand coordinated international efforts – in standards, strategies, and technologies.

7.2 Digitalization as a Key Enabler

The digital transformation of plastics recycling is opening up major opportunities to enhance efficiency, quality, and transparency. Key developments discussed by industry experts include the following:

AI-Based Sorting and Recycling

Artificial intelligence is already transforming material separation. Companies like Tomra and ZenRobotics have demonstrated that AI-supported sorting can achieve significantly higher purity levels, enabling high-quality recycled materials even from complex waste streams (see also Chapter 2).

Digital Product Passports (DPP) and Digital Twins

The development of digital product passports offers a critical tool for tracking material composition, recyclability, and environmental impact throughout a product's life. When linked with digital twins, which create virtual representations of products, recyclability tracking and predictive maintenance can become even more powerful.

Blockchain for Traceability

Blockchain technology holds great promise for securing material flows and ensuring transparency. Early pilots, such as those by Circularise, show how blockchain could verify recycling content and product history, though scalability challenges remain.

Online Marketplaces for Recyclates

Platforms like Cirplus and RecyClass simplify access to high-quality recyclates, promoting transparency, setting clear quality standards, and helping stabilize demand as recycled content requirements increase.

Recycling Apps

Apps such as Recyclops, Bower, and TrashScan are reshaping consumer engagement, offering education, participation incentives, and easier recycling navigation.

Material Innovation through Digital Tools

Simulation software, big data, and AI are driving the creation of new polymer materials optimized for recyclability or biodegradability, as seen in companies like Itero Technologies, BASF, and Covestro.

Standards and Norms

Efforts such as EN 18065 and the DIN EN 1534X series are laying the foundation for consistent recyclate classification and easier integration into industrial processes (see also Section 3.3).

Data Analysis and Optimization

While major enterprises already leverage platforms like SAP and IBM for carbon footprint tracking and material flow optimization, broader industry adoption is still evolving.

Design for Recycling

Design-for-recycling strategies have become common in packaging development and are spreading across industries. However, further adaptation is necessary to align complex products with end-of-life requirements

Digitalization will not solve all challenges alone, but it will be a critical enabler for achieving transparency, traceability, and better material performance in the circular economy.

7.3 Remaining Global Challenges and the Need for Systemic Change

Despite impressive technological advancements, the global plastics recycling ecosystem still faces significant hurdles.

Even high-income countries such as the United States lag behind global leaders like Germany in waste collection efficiency, sorting infrastructure, and integration into circular material flows. Historical reliance on exporting plastic waste has exposed the vulnerability of systems without robust domestic waste management strategies.

Meanwhile, developing countries often lack even basic waste collection services, leading to uncontrolled plastic leakage into the environment and oceans.

Without a dramatic global improvement in waste collection, sorting, and domestic processing, the circular economy cannot succeed.

Even the most advanced recycling technologies depend on a consistent and sufficient supply of recoverable materials. Without proper collection systems, there will simply not be enough recyclate available to meet regulatory targets, market demands, or environmental goals.

The shift toward circularity requires a systemic transformation – spanning technology, policy, infrastructure investment, and cross-border cooperation.

7.4 Shaping the Future: A Circular Economy Built on Collaboration, Innovation, and Transparency

The next decade offers a unique opportunity to align environmental goals with economic realities. Future policy frameworks must become more integrated, science-based, and innovation-driven.

Emerging instruments like Digital Product Passports and blockchain will strengthen material traceability, but only if accompanied by international harmonization and practical market incentives.

Eco-modulation of EPR fees, mandatory recycled content standards, and design-for-circularity regulations will be critical levers for accelerating change.

However, regulation alone is not enough:

- Industries must commit to quality and traceability over volume alone.

- Governments must build efficient, inclusive waste management systems.

- Consumers must be empowered to participate meaningfully.

Global collaboration will be the cornerstone: harmonizing definitions, supporting infrastructure in developing regions, and establishing shared standards for product design, material tracking, and recycling performance.

The transformation of the plastics economy is no longer optional – it is a necessity.

But success will not be measured by the tonnage of material collected; it will be defined by the quality, traceability, and systemic integration of material flows. Achieving true circularity demands an honest, pragmatic, and collaborative approach. It demands innovation not only in technology but in policy, business models, and international cooperation. It demands that we design, manage, and recover materials with the future – not just the present – in mind.

By putting quality first, embracing transparency, and working together across borders, industries, and societies, we can finally move beyond today's linear plastics economy, and build a resilient, sustainable system that serves both people and planet.

8

Appendix

This appendix includes the content of Chapters 4, 6, and 8 of the 2nd edition of this book, contained in Section 8.1, Section 8.2, and Section 8.3, respectively. Cross-references have been updated for the new numbering.

8.1 Economic Analysis of Plastic Waste Handling

The first step to improve plastic waste handling is analyzing the procedures involved from an economic point of view. In order to do so, we will first review the *fundamentals of economic efficiency*. Afterwards, using these principles, we will analyze the economic efficiency of landfilling, incineration with energy recovery, and recycling, and then compare them.

8.1.1 Fundamentals of Economic Analysis

8.1.1.1 Economic Efficiency Calculation

In theory and practice, a variety of techniques to evaluate and compare economic efficiency exist. They are derived from classic capital budgeting, which considers the advantage of an investment with known acquisition costs and was designed for helping select the best long-term investments for financing [1, 2].

Nowadays, capital budgeting also includes the valuation of companies and processes, via the so-called *economic efficiency calculation*. Its purpose is to measure the profitability of a company or process in order to derive an asking price for that company or process. It also enables the comparison of the efficiency of different processes with each other [1]. Therefore, the economic efficiency calculation can be used to compare landfilling, incineration with energy recovery, and recycling from an economic point of view.

In economic theory and entrepreneurial practice, there are two main approaches to making this calculation:

- *Static methods* determine the economic efficiency under constant framework conditions and explicitly consider only one time period, which is assumed to be representative for all such time periods. A typical time period under review is a year. The data for this time period is obtained by considering the whole planning period (in general the expected lifetime of an investment) and finally deriving the average cost and profit data for the time period under review [1, 3].

- *Dynamic methods* also analyze the influence of the time of investment on the economic efficiency. By discounting payments and earnings using identical benchmarks via compound interest calculations, the impact of both the amount and the time of the investment on the profitability are considered. These methods are in general closer to reality, which is a clear advantage over static methods. If, for example, an investment in a new machine needs to be analyzed, dynamic methods enable one to include factors such as learning effects and decreasing manufacturing costs, which have a big impact on the economic efficiency [1, 4, 5, 6].

The challenge of dynamic methods is the data collection and the uncertainty of future payments, earnings, and market behavior (e. g., inflation), which can lead to inaccuracies in the results. For this reason, static methods are still more frequently used. In practice, static and dynamic methods are often used in combination, since companies do not want to rely on only one of these methods alone [1].

In the following, static methods will be used to analyze economic efficiencies of plastic waste handling.

8.1.1.2 Static Economic Efficiency Calculation

As mentioned in Section 8.1.1.1, static economic evaluations consider one average period (e. g., a year), which is assumed to be representative for the lifetime of the process. These methods focus on a pure financial measure of the profitability, which can be thought of in two ways:

- *Absolute profitability:* Investing in and/or running a business or process is economically better than not running it.

- *Relative profitability:* Investing in and/or running business A or process A is economically better than investing in and/or running business B or process B.

Static methods assume constant payments and earnings and do not take into account the changing value of cash over time. There are different static economic calculation methods, from which the profit comparison method is the most used one. Thus, this method will be further considered [1, 3, 7].

8.1.1.3 Profit Comparison Method

The *profit comparison method (PCM)* computes the profit of plastic waste handling processes by subtracting the total costs of these processes from their total revenues. The two types of profitability considered are:

- *Absolute profitability:* the profit of a waste handling procedure

- *Relative profitability:* the profit of a particular waste handling procedure as compared to the profit of another waste handling procedure

The costs included in the calculation are [3, 5]:

- *Initial investment cost*

- *Capital cost*: depreciation and interest

- *Fixed costs*: lease rental charges and/or land costs, fixed salaries, and setup costs

- *Variable costs*: costs of material and tools; direct labor costs; energy, electricity, and water costs; maintenance costs; and taxes

For plastic waste handling, where the revenue comes from depends on the procedure: waste-to-energy (WTE) plants derive revenue from selling electricity produced, and recycling plants from selling the recycled pellets or plastic products.

After summing up all costs and revenues for each plastic waste handling procedure, the average cash flows for one representative time period or unit need to be determined to estimate the profit of each procedure. Here the absolute and relative profitability will be determined for handling of **1 ton (t) of plastic waste** [3].

8.1.2 Economic Analysis of Landfilling

Landfills are required to comply with federal regulations, which include among other things location restrictions, design specifications, operating standards, and closure requirements. These regulations have a great influence on the cost to build, operate, and finally close landfills [8, 9].

Besides regulatory issues, *landfill costs* are site specific; thus, they vary based on factors such as terrain, soil type, climate, site restrictions, and amount of waste disposed. For economic analyses, the costs of landfills are mainly grouped into four categories [9, 10, 11, 12, 13]:

- Construction costs

- Operations costs

- Closure costs

- Postclosure and maintenance costs

To analyze the profit of landfilling, a large landfill with a footprint of 365.76 m × 365.76 m (1,200 ft × 1,200 ft, approximately 33 acres) will be analyzed. The trend in the United States is towards large landfills. The average landfill has a yearly disposal of 70,000 t (for a total of around 140 Mt per year on 1,269 landfills). The landfill considered here has a yearly disposal of 200,000 t, or 500 to 600 t daily.

Although most of the literature values were based on the U.S. customary or imperial system (feet [ft], acres), Section 8.3 includes calculations both for the imperial and metric system. Table 8.1 summarizes important assumptions that are required for the economic analysis [11, 12, 13].

Table 8.1 Assumptions for Economic Analysis of Landfill

Lifetime [years]	11	
Postclosure care period [years]	30	
Final surface grades [acres] | [hectares (ha)]	34.00	13.76
Bottom of landfill [acres] | [hectares (ha)]	33.50	13.56
Disposal capacity constructed acre [yd^3/acre] | [m^2/m^3]	60,000	19.39
Total capacity [yd^3] | [m^3]	4,000,000	3,058,104
In-place density of waste [t/yd^3] | [t/m^3]	0.55	0.72
Weight capacity per acre [t/acre] | [t/m^2]	33,000	49.42
Waste per day [t]	500–600	
Waste per year [t]	200,000	
Total weight capacity [t]	2,227,500	
Total overall weight of landfill (11 years) [t]	2,200,000	

To calculate the profit of landfilling, the costs must be considered first. Therefore, the total overall costs of the 13.35 ha (33 acres) landfill will be summarized and finally divided by total weight of municipal solid waste (MSW) disposed over the landfill's lifetime (2,200,000 t).

The first cost category is *construction*, which is divided into that for the support facility and that for the liner construction. Based on the 365.76 m × 365.76 m (1,200 ft × 1,200 ft) footprint, the landfill's access road has a length of approximately 381.00 m (1,250 ft), in total 1,524.00 m (5,000 ft). Including possible setbacks for the landfill from the property line required by the state regulatory agency and additional area required for other facilities, the security fence has a length of 1,828.82 m (6,000 ft). The costs per foot or per meter of road and fence are given in Section 8.3 in Figure 8.13 (imperial) and Figure 8.14 (metric), respectively. The space and costs for support buildings, including office buildings, maintenance buildings, shacks, and tool sheds, are found in Figure 8.13 (imperial) and Figure 8.14 (metric). In addition, every landfill plant needs one modular truck scale and an associated computer system [13].

The bottom of the landfill with an area of 13.56 ha (33.5 acres) needs a liner and leachate system. After performing a site survey, the bottom area first needs to be cleared and grubbed. Once completed, liner construction grades and elevations are established by excavation. It is assumed that the landfill has a minimal structural fill berm constructed along the landfill's perimeter to provide anchoring for the liner elements and structural toe stability for the final waste slope. After establishing the base grades, the landfill's liner and leachate management system, including clay liner, geomembrane, geocomposite, and granulator soil, is constructed. Besides these physical acts of construction and installation, management and quality oversight are required, typically done by independent third-party consultants. The detailed statement of costs is also found in Figure 8.13 (imperial) and Figure 8.14 (metric).

On average, the overall *construction costs* for the completed landfill total **$21,582,950**, as seen in Table 8.2 [11, 13].

Table 8.2 Construction Costs of Landfill

	Minimum	Maximum	Average
Support facility construction [$]	1,175,300	1,790,600	1,482,950
Liner construction [$]	11,256,000	28,944,000	20,100,000
Total construction costs [$]	**12,431,300**	**30,734,600**	**21,582,950**

Operating costs include staffing, utilities, equipment operations, leachate disposal and treatment, scale operations, paperwork, record keeping, billing, engineering staffing, environmental monitoring, and daily cover applications. Assuming waste processing of 200,000 t per year, which is equivalent to waste receipt of 500 to 600 t per day, this landfill would at minimum require a front-end loader for onsite hauling of bulk material and small construction tasks (e. g., CAT 950H), a bulldozer equipped with a trash rack to spread dirt and waste (e. g., CAT D7), and a steel-wheeled compactor to compact the waste and achieve maximum possible in-place density (e. g., CAT 826G). Additional equipment such as water spray trucks (for holding dust down), a scraper, a backhoe, several pickup trucks, and a road grader would also be needed [14, 15, 16].

The salaries of all employees are based on average U.S. values. Summarizing all operations costs, the yearly operations costs amount to **$1,578,300**, or **$17,361,300** after 11 years, as presented in Table 8.3. Detailed operations cost calculations are shown in Figure 8.15 in Section 8.3 [11, 13, 17, 18, 19, 20, 21].

Table 8.3 Operations Costs of Landfill

Cost Factor	Cost [$]
Operations [$]	1,368,300
Leachate collection and treatment [$]	30,000
Environmental sampling and monitoring [$]	60,000
Engineering services [$]	120,000
Total operations costs per year [$]	**1,578,300**
Total operations costs of landfill (11 years) [$]	**17,361,300**

After use for disposing of waste for 11 years, the landfill must be closed. Due to sloping, the final surface grades needing capping and covering are approximately 13.76 ha (34 acres). The closure process starts with surveying the surface to receive a final cap and cover. Once this step is done, construction of the final cap and cover begin, which includes a geomembrane cap and vegetative soil, seed, mulch, and fertilizer. To prevent air and soil pollution, run-off and gas control systems are installed. Besides these physical acts of construction and installation, management and quality oversight are required. A detailed cost breakdown is shown in Figure 8.16 (imperial) and Figure 8.17 (metric) in Section 8.3. Summarizing the mentioned categories, *closure costs* of the considered landfill range from **$7,718,000** to **$11,084,000**, with an average total cost of **$9,401,000** [11, 12].

Once a landfill is closed, RCRA prescribes the care for and maintenance of landfills for a minimum of 30 years. *Postclosure and maintenance costs* are mainly divided into site security maintenance, landfill cover, and mechanical systems maintenance, monitoring wells and gas probes, and environmental monitoring. For 1 year, these costs range between $72,182 and $100,232 for the complete landfill, which is shown in Figure 8.18 (imperial) and Figure 8.19 (metric) in Section 8.3. Over 30 years, the complete postclosure and maintenance costs add up to **$2,586,210**.

Table 8.4 Total Costs of Landfill

Category	Minimum	Maximum	Average
Construction costs [$]	12,431,300	30,734,600	21,582,950
Operations costs [$]	17,361,300	17,361,300	17,361,300
Closure costs [$]	7,718,000	11,084,000	9,401,000
Postclosure/maintenance costs [$]	2,165,460	3,006,960	2,586,210
Total costs [$]	**39,676,060**	**62,186,860**	**50,931,460**
Total costs per ton [$/t]	**18.03**	**28.27**	**23.15**

To finally calculate the cost per ton of plastic waste, the total costs of all categories are summed up and divided by the total waste disposed of in the landfill (2,200,000 t).

As shown in Table 8.4, the total average cost of landfilling 1 t of plastic waste is **$23.15**.

The revenue from landfilling 1 t of plastic waste is **$0.00**. Nevertheless, it is possible for landfill owners to receive revenue by selling carbon credits, selling electricity generated from landfill gas (LFG) to the local power grid, or selling LFG to a direct end user or pipeline. However, the revenue received represents a negligible percentage of the operating costs of a landfill and is therefore set to $0.00 [9, 22].

Applying the profit comparison method, the last step is to calculate the overall profit of landfilling 1 t of plastic waste.

Table 8.5 Total Profit per Ton of Plastic Waste for Landfill

	Minimum	Maximum	Average
Revenues per ton [$/t]	0.00		
Costs per ton [$/t]	18.03	28.27	23.15
Profit per ton [$/t]	**–18.03**	**–28.27**	**–23.15**

Subtracting the cost per ton from the revenue per ton, the average profit of landfilling plastic waste is –**$23.15 per ton**. This means landfilling is not absolutely profitable (profit < $0.00).

The figure of –$23.15 per ton is the costs for a big landfill. The bigger the landfill, the smaller the relative costs. If the average waste per day were for example only 450 t, but the landfill is constructed for 550 t per day, the yearly waste would be around 165,000 t and after 11 years only 1,815,000 t. The landfill would have to be closed due to national regulations; the total costs would be the same and the costs per ton would be on average **$28.27**.

Another big cost factor that is not included in the calculations is the costs for cleaning up landfills in cases of accidents. Since most of the landfills are not cleaned up afterwards due to high costs, the experiences with these costs are comparably low. But some examples can be found and allow us to estimate this cost. In 2008, a dike failed at the Tennessee Valley Authority's Kingston Fossil Plant. A total of 5.4 million yd³ of coal ash cascaded into the Emory and Clinch Rivers and smothered about 300 acres of land. Up until 2015, about **$1.2 billion** were spent to clean up the area, and more will probably follow. This is about 19 times more than the maximum costs for the landfill we are considering here [23].

In the future, costs for landfilling will probably increase substantially. Since land is becoming a more and more scarce resource, the costs for land and construction are likely to rise. This trend can be seen in the average *tipping fee*, also called the gate fee, of landfills in the United States. Between 1985 and 2018, the average tipping fee increased from **$8.22** to **$55.00** per ton of waste. The increase was likely caused by states implementing RCRA Subtitle D regulations or their state's equivalent. Furthermore, China's import ban for plastic waste leads to an increased average tipping fee [13, 22, 24, 25, 60].

8.1.3 Economic Analysis of Incineration with Energy Recovery (Waste-to-Energy Facilities)

The most widespread type of waste-to-energy plants in the United States are mass-burn facilities (64 out of 84), and therefore we use an example of one for this profitability analysis [26].

The profit analysis of mass-burn facilities is mainly split into three categories [27]:

- Investment costs
- Operating and maintenance costs
- Revenues from sale of heat and electricity generated

The actual profit for a waste incineration plant depends on a wide range of factors, especially the capacity of the plant and the type of energy production. The facility considered here is a co-generation facility and produces both heat and electricity. Since new WTE facilities have an expected lifetime of 35 years, this period was assumed for the following analysis. As for the landfill analysis (Section 8.1.2), the complete profit of plastic combustion over the facility's lifetime is summed up and divided by the total plastic waste burned (3,500,000 t). The important general assumptions of this analysis are shown in Table 8.6, and in more detail in Figure 8.20 in Section 8.3 [28, 29].

Table 8.6 Assumptions for Economic Analysis of Waste-to-Energy (Mass-Burn) Plant

Lifetime [years]	35
Yearly waste capacity [t]	100,000
Total waste capacity (35 years) [t]	3,500,000
Steam generator efficiency [%]	80
Percentage of plastic in MSW [%]	12.8
Dollar–Euro exchange rate [$/€]	1.2125
Purchasing power parity (PPP) conversion factor of Croatia	0.6

Data for *investment costs* was taken from a waste-to-energy plant in Croatia, since this was the only comprehensible data available. To realistically transfer this data to the U.S. market, it first needed to be converted from Dollars to Euros at an average for the last 20 years of **$1.21** per Euro [30].

Furthermore, the prices were adjusted using the *purchasing power parity (PPP)* conversion factor, which represents the number of units of a country's currency required to buy the same amount of goods and services in that country as a U.S. Dollar would buy in the United States. The PPP conversion factor of Croatia was consistently 0.6 over the last 15 years and therefore is used for these calculations [31].

Investment costs vary substantially with respect to several factors, including the design of the WTE plant, its capacity, and the local infrastructure. Construction of a road, a weighing area, and waste reception storage is necessary. The combustion chamber-steam generator system includes a system for feeding waste into the combustion chamber, an air supply for the combustion chamber, a combustion chamber, ash removal and storage, flue gas channels, and a steam generator with steam output. The water and steam system consists of a water treatment facility, an air-cooled condenser, and a condensation turbine. Flue gas cleaning, which represents an important part of the overall waste combustion process, is ensured by a flue gas cleaning system involving a semi-dry treatment, a bag filter, and a selective noncatalytic reduction (SNCR) system. Additional initial costs are design, construction, electromechanical installations, and other smaller costs. Summarizing all these costs, the total investment for the WTE plant amounts to **$113,781,924**, as shown in Table 8.7 and in more detail in Figure 8.21 in Section 8.3 [29].

Table 8.7 Investment Costs of Waste-to-Energy Plant

Cost Factor	Cost [$]
Infrastructure and waste storage [$]	9,296,569
Combustion system and steam generator [$]	39,409,370
Water and steam system [$]	16,167,946
Design [$]	4,041,986
Construction [$]	14,146,953
Electromechanical installations [$]	10,104,966
Other investment costs [$]	12,125,960
Flue gas cleaning system [$]	8,488,172
Total investment costs [$]	**113,781,924**

Operating and maintenance costs are analyzed annually and divided into personal salaries, operation and maintenance costs as well as heat and electricity costs. The plant runs three shifts daily. Per shift, 15 workers, 5 environmental engineers, 3 maintenance engineers, and 1 facility manager are required. Salaries used for the calculations are average U.S. salaries. The annual salaries of all personnel amount to $3,570,000. Maintenance costs of the combustion system are proportional to the waste flow and estimated at 3% of total investment costs. Additional annual machine and emission costs are process water, natural gas, bottom ash disposal, and bag filter residues costs. Total annual machine and emission costs are $10,190,980. Assuming electricity consumption of $0.1\,MW/t_{waste}$ with a price of $0.1027 per kWh and a heat consumption of $0.05\,MW/t_{waste}$ with a price of $0.0011 per kWh, the annual heat and electricity costs are $105,454. Operating and maintenance costs add up to $13,866,434 per year and $485,325,214 after 35 years, as presented in Table 8.8. A more detailed calculation is provided in Figure 8.22 in Section 8.3 [21, 29, 32, 33, 34, 35].

Table 8.8 Operating and Maintenance Costs of Waste-to-Energy Plant

Cost Factor	Cost [$]
Personnel salaries per year [$]	3,570,000
Machine and emission costs per year [$]	10,190,980
Heat and electricity costs per year [$]	105,454
Yearly operating and maintenance costs [$]	**13,866,434**
Overall operating and maintenance costs [$]	**485,325,214**

The revenues from plastic combustion come from selling the electricity and heat produced by these plants. These revenues depend on the average lower heating factor of burned material. Compared to general MSW, the average lower heating value of plastics is high. Based on the composition of plastics in the MSW stream, the lower heating value of plastics in waste is **36.16 MJ/kg** (detailed calculations are found in Figure 8.23 in Section 8.3). The general MSW has an average lower heating value of only **14.98 MJ/kg** (detailed calculations are found in Figure 8.24 in Section 8.3). All values and assumptions necessary for the revenue calculations are presented in Table 8.9 [36, 37].

Assuming a percentage of plastics in MSW of 13.2%, 13,200 t of plastic waste (out of 100,000 t) are incinerated annually. Theoretically, the energy produced through plastics combustion is therefore 462,833,408 MJ per year, 75% of which produces heat and 25% of which produces electricity. Assuming a steam generator efficiency of 80%, the electricity produced yearly amounts to 32,141,209 kWh, and the heat produced yearly to 96,423,627 kWh [38].

Table 8.9 Assumptions for Revenue Calculations from Sale of Energy Generated in Waste-to-Energy Plant

Percentage heat of produced energy [%]	75
Percentage electricity of produced energy [%]	25
Energy per cubic meter in gas [kWh/m^3]	12.5
Average lower heating value MSW [MJ/kg]	14.98
Average lower heating value plastic [MJ/kg]	36.16
Costs/Revenue per kWh electricity [$/kWh]	0.1027
Cost/Revenue per cubic ft gas [$/ft^3]	0.0039
Cost/Revenue per cubic meter gas [$/m^3]	0.1377

The average price of electricity for customers in the United States in December 2019 was $0.1027 per kWh. If WTE plants produce 32,141,209 kWh/year, annual revenues from selling electricity generated by plastics combustion are $2,561,011. Supposing an energy content of 12.5 kWh per cubic meter gas, the generated heat would produce 6,171,112 m^3 of gas per year. With an average gas price of $0.1377 per m^3, annual revenues of selling gas produced by plastics combustion are $849,931 [35].

Summing up the revenues of the sale of both the electricity and heat and multiplying it by the plant lifetime of 35 years, total overall revenues of incinerating plastics add up to **$119,383,002**, as shown in Table 8.10.

Table 8.10 Total Revenue from Sale of Energy Generated in Waste-to-Energy Plant

Revenue Type	Revenue [$]
Yearly revenue for electricity [$]	2,561,011
Yearly revenue for heat [$]	849,931
Yearly revenue of plastic waste combustion [$]	**3,410,942**
Total revenue of plastic waste combustion [$]	**119,383,002**

Finally, to calculate the profit from combustion of 1 t of plastic waste, the costs need to be subtracted from the revenues. Therefore, the investment, operations, and maintenance costs are added and divided by total waste burned over the plant's lifetime (3,500,000 t). Then the total revenues of plastic waste combustion are divided by the total amount of plastic burned over 35 years (448,000 t). These calculations result in costs of $171.17 and revenues of $266.48 per ton of plastic waste, which leads to a profit of **$95.31** per ton of plastic waste burned, as shown in Table 8.11.

Table 8.11 Profit per Ton from Combustion of Plastic Waste

Revenues per ton of plastic waste burned [$/t]	266.48
Costs per ton of plastic waste burned [$/t]	171.71
Profit per ton of plastic waste burned [$/t]	**95.31**

The profit of **$95.31** per ton of plastic waste burned is an absolute profit (profit > $0.00). As already mentioned, this value can only be reached theoretically. The calculation assumed that no plastic waste was recycled or treated any other way. Furthermore, in a real scenario, plastic waste would not be combusted separately from other waste. But due to its high average lower heating value, plastic waste combustion results in high energy production and is absolutely profitable.

To review and check the WTE-plant analysis, the profit of combustion of 1 t of MSW was calculated and compared to the average tipping fee of combustion of 1 t of waste in the United States. The costs of burning 1 t of MSW are equivalent to the costs of incinerating 1 t of plastic waste ($171.71). To determine revenues of MSW combustion, an average lower heating value of 14.98 MJ/kg was calculated (see Figure 8.24 in Section 8.3). This resulted in a yearly revenue (from heat and gas) of $10,808,957, or **$378,313,500** after 35 years. Dividing the total revenues by the total incinerated MSW over the plant's lifetime, the revenue from combustion of 1 ton of MSW is $110.38.

Subtracting costs from revenues, the total profit of MSW combustion is –$60.80 per ton of MSW.

Table 8.12 Profit per Ton from Combustion of Municipal Solid Waste (MSW)

Revenues per ton of MSW [$/t]	110.38
Costs per ton of MSW [$/t]	171.71
Profit per ton of MSW [$/t]	**–60.80**

The average tipping fee for combustion of 1 t of MSW in the United States is **$64.96 per ton** (Figure 8.25) and represents approximately the costs of waste combustion. This fee only differs by **$4.16** (6.4%) from the calculated value and therefore indicates reasonable assumptions and calculations of the previous WTE plant analysis [39, 40, 41].

A factor not being included in the profit analysis is the case of rebuilding a plant: a new plant could be built on the site of the existing WTE plant, which reduces the capital cost for land in the new facility to zero. This is a big economical advantage, especially compared to landfill plants, the land from which cannot be used again after closure for a long postclosure period [42].

8.1.4 Economic Analysis of Plastics Recycling

The third plastic handling method is *recycling*. Figure 8.1 shows the plastics recycling process beginning and ending with the consumer.

After plastic is thrown away and collected within the single recycling stream, material recovery facilities (MRF) sort the material into the seven plastic recycling groups [43].

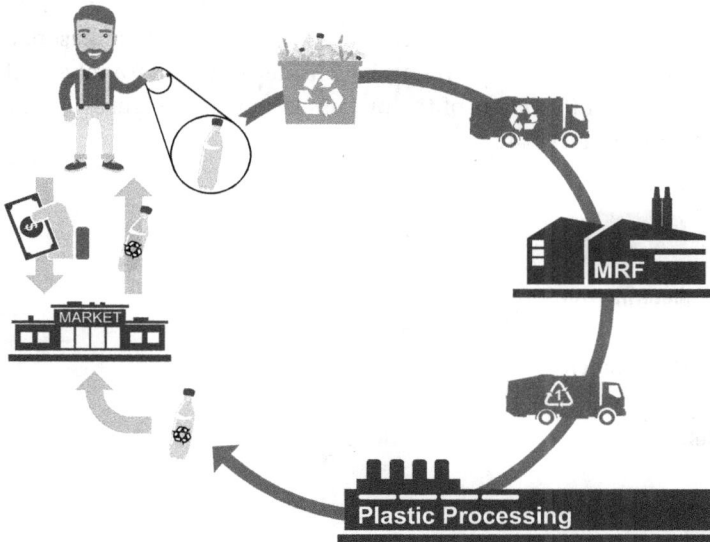

Figure 8.1
Overview of the plastics recycling process

The higher the resin number, the harder the recycling of that plastic. In the MRF, plastics are sorted into three groups: first, number 2 plastic (HDPE) is sorted out manually and next, number 1 plastic (polyethylene terephthalate [PET]) is optically separated from numbers 3 to 7 plastics with an automated IR detector. Due to its good recycling properties, number 1 plastic is the most valuable and at present the only plastic in the United States that is recycled, reprocessed, and resold as pure material in significant quantities. Consequently, the economic analysis of the plastic recycling process in this book will be based on PET recycling.

There are three steps to the economic analysis of the profitability of PET recycling that will be analyzed in the following:

▪ The cost of material recovery in the MRF (Section 8.1.4.1)

▪ The cost of plastic reprocessing (Section 8.1.4.2)

▪ The revenue derived from selling recycled PET (Section 8.1.4.3)

8.1.4.1 Materials Recovery Facility Costs

Plastic waste is collected by dump trucks together with all the other recyclable materials such as metal, paper, and glass waste and brought to a *materials recovery facility (MRF)*. A MRF is a processing facility where waste is sorted and prepared for additional processing. Configurations of the MRF processing line vary depending upon how waste is collected and received at MRFs. The four different ways of collecting waste are source-separated, dual-stream, single-stream, and mixed-waste collection. Due to advances in automated processing equipment, less upfront separation is required; thus, many source-separated MRFs gave way to dual-stream MRFs, which in turn are being replaced by single-stream MRFs in some locations. For this reason, a semi-automated MRF for single-stream recycling is considered for the analysis of plastics recycling in this book. A flowchart of the process is shown in Figure 8.2. The solid lines represent the flow of PET [44].

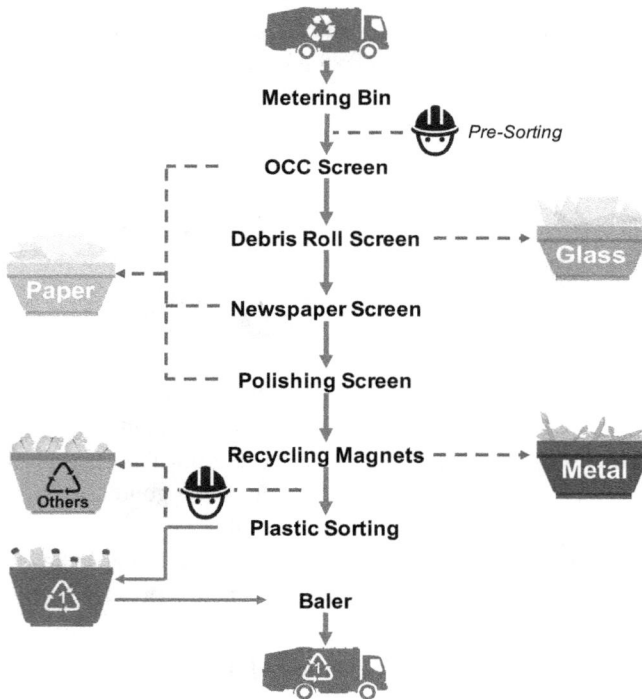

Figure 8.2
Schematic of materials recovery facility (MRF). OCC screen, old corrugated cardboard screen

As trucks arrive at the MRF, they unload recyclable waste onto the tipping floor, where the material is stored until processing. Usually, the tipping floor is sized to hold at least two days of incoming waste to allow a buffer against unscheduled equipment downtime and to provide sufficient material for the MRF to operate during an extra shift. The dumped waste is grabbed by a loader to fill a *metering bin;* this serves to produce a constant flow. This constant feed rate prevents surges, allows for more efficient manual sorting in the presort area, and maximizes the efficiency of automated equipment encountered later in the processing line. During the manual presorting, large contaminants, bulky recyclables, and items that could damage downstream sorting equipment or pose a threat to personnel are removed [44, 45].

As material leaves the presort station, it is transferred to the next machine, the so-called *old corrugated cardboard (OCC) screen*. It is composed of several rotating star-shaped disks, which separate the OCC from all the other materials by using size and dimension. All non-OCC materials fall through the disks onto the next screen. Below the OCC screen is a *debris roll screen*, which acts as a glass breaker. It breaks glass into quarter-sized pieces, which then fall through the debris roll screen and go into the glass trailer for transportation to a glass recycler.

After OCC and glass are removed, residual waste is routed to a *newspaper screen*, which uses the same concept as the OCC screen. The difference between these screens is the distance between the disks. Disks of the newspaper screen are closer together, so that the newspaper material stays on the disks and the smaller materials fall through. The *polishing screen* has even closer disks, which separate remaining paper waste from other materials [44, 45].

OCC, newspapers, and other paper materials are finally routed to a paper platform, where the paper waste is checked manually and contaminants, which "acted" like paper when passing through the three paper screens, are removed from the paper waste before paper is sent out to a paper recycler [44, 45].

A second system, called an *eddy current separator*, induces an electrical current in conducting metals such as aluminum. This in turn induces a magnetic field in these materials, repelling the aluminum and thus removing it from the waste stream [44, 45].

The only materials remaining in the system are numbers 1 to 7 plastics. In a first step, number 2 plastics (HDPE) are taken out manually. Afterwards, the residual plastics are sorted by an *optical sorter*. This machine shoots a ray of light into the plastic and analyzes how the light reflects back to determine if the item is a number 1 (PET) or numbers 3 to 7 plastics. The optical system shoots a controlled blast of air on the item, which directs it to either the number 1 or the numbers 3 to 7 plastic storage area. Non-plastic materials do not receive any air shot and fall down to the residue conveyor, ending in the trash [44, 45].

Sorted PET is then in the interim stored in a separate material bunker until the loose material is compacted by a *baler* into a large, bricklike bale to prepare the material for shipping to a plastic processing company [44, 45].

The costs of the MRF process are split into two categories, investments and operating costs. As in the landfill and WTE facility analysis, all costs over the plant's lifetime are summarized first and afterwards converted to a cost per 1 t of plastic waste processed.

The assumptions for the economic analysis are presented in Table 8.13 and in greater detail in Figure 8.26 and Figure 8.27 in Section 8.3 [46, 47].

Table 8.13 Assumptions for Economic Analysis of Materials Recovery Facility

Lifetime [years]	10
Yearly working hours [h]	4,160
Waste handled per hour [t]	30
Yearly waste handling [t]	120,000
Total waste handling (10 years) [t]	1,200,000
Residues rate [%]	2
Correctness of manual sorting [%]	91

Since the factory runs 52 weeks per year, 5 days per week in 2 shifts with an effective working time of 8 hours each shift, the yearly working hours add up to 4,160 hours. Since 30 tons are handled per hour (restricted by the capacity of the metering bin), around 120,000 tons of waste is handled per year, resulting in a total waste handling of 1,200,000 tons of waste after 10 years [47].

Investment costs for the MRF are divided into buildings and site, machine, and equipment costs. For an 80,000 ft^2 large area, initial costs for the land and site work are $1,395,000, and buildings costs for the scale house and MRF building are $9,200,000. Construction, planning, and surveying costs amount to $3,500,000, so the total investment costs for the buildings and site add up to $14,095,000. The investment costs of all machines for the MRF (Figure 8.2) amount to $2,163,000. Additional equipment are conveyors, rolling stock (e. g., front-end loaders and forklifts), and collection cars, the costs of which add up to $1,400,000. Summing up the investment costs from these three categories, the total investment costs of an MRF are **$17,658,000**, as shown in Table 8.14 and in more detail in Figure 8.28 in Section 8.3 [46, 47, 48].

Table 8.14 Total Investment Costs of Materials Recovery Facility

Cost Factor	Cost [$]
Building and site investment costs [$]	14,095,000
Machine investments costs [$]	2,163,000
Additional equipment investment costs [$]	1,400,000
Total investment costs [$]	**17,658,000**

Operating and maintenance costs of the MRF are divided into seven different categories. The yearly operating and maintenance costs are $8,451,851, so overall operating and maintenance costs of $84,518,517 after 10 years, as shown in Table 8.15 and in more detail in Figure 8.29 in Section 8.3.

Table 8.15 Total Operating and Maintenance (O&M) Costs of Materials Recovery Facility

Cost Factor	Cost [$]
Personnel salaries per year [$]	2,552,000
Facility costs per year [$]	343,000
Machine O&M costs per year [$]	130,602
Conveyor O&M costs per year [$]	12,392
Rolling stock O&M costs per year [$]	897,056
Residues costs [$]	96,000
Transportation and collection costs [$]	4,420,800
Yearly O&M costs [$]	**8,451,851**
Overall O&M costs (10 years) [$]	**84,518,517**

The largest group of personnel is the sorters for presorting, manual HDPE sorting, and quality control. A detailed breakdown of personnel is provided in Figure 8.29 in Section 8.3. The yearly salaries amount to $2,552,000. The salaries used for the calculation are average yearly U.S. salaries [18, 19, 20, 21, 48].

Facility costs include consumables, insurance, administration, and baling wire and amount yearly to $343,000. Operating and maintenance costs of machines as well as conveyors involve yearly maintenance and electricity consumption ($0.1027 per kWh) and add up to yearly $142,995. The rolling stock has a high consumption of diesel, so the yearly costs are $897,056. Assuming that 2% of the waste is residues and that all residues are landfilled, yearly costs for residues are $96,000. The largest part of operating and maintenance costs is generated by transportation and collection: assuming transportation and collection costs of $36.84 per ton, yearly costs for 120,000 t are $4,420,800.

Summarizing the investment costs and overall operating and maintenance costs, the total MRF costs are **$102,176,517**, or **$85.15** per ton of plastic waste (Table 8.16).

Table 8.16 Total Costs of Materials Recovery Facility

Cost Factor	Cost [$]
Investments costs [$]	17,658,000
Operating and maintenance costs [$]	84,518,517
Total costs (10 years—1,200,000 t) [$]	**102,176,517**
Costs per ton of plastic waste [$/t]	**85.15**

8.1.4.2 Plastic Reprocessing Costs

After PET is baled in the MRF, the bales are transported to a *plastic reprocessing facility*, where they are further treated, as schematically presented in Figure 8.3.

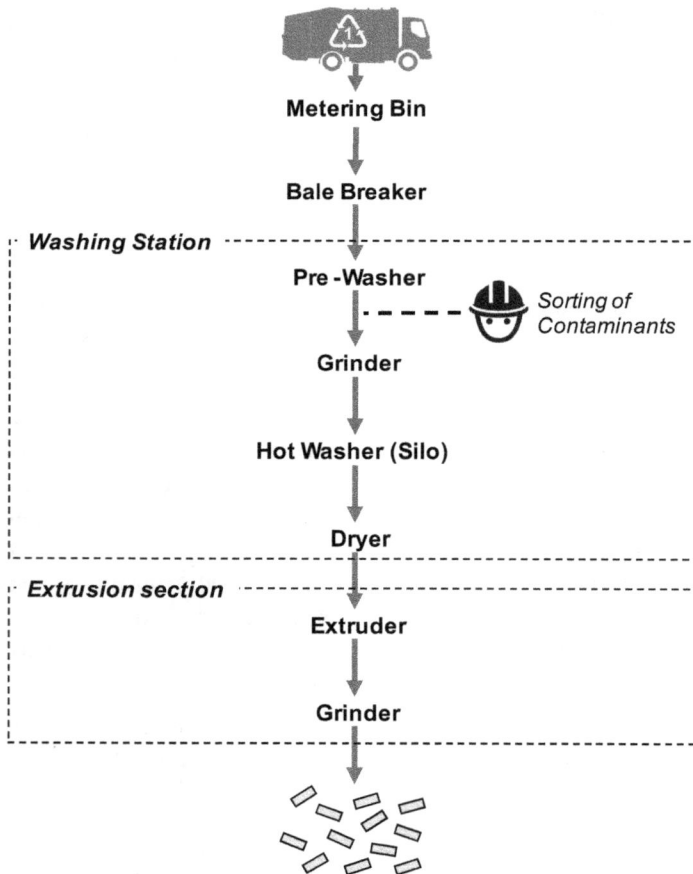

Metering Bin

Bale Breaker

Washing Station

Pre -Washer

Sorting of Contaminants

Grinder

Hot Washer (Silo)

Dryer

Extrusion section

Extruder

Grinder

Figure 8.3
Schematic of plastic reprocessing facility

From the tipping floor, PET bales are grabbed by a loader and laid into a metering bin, which constantly meters the plastic waste into a bale breaker. The bale breaker dismembers the PET bales into individual free flowing items (e. g., food containers and bottles) [49, 50].

The individual items are conveyed to a washing station. After a short prewashing to remove labels and dirt from the outside of the items and a manual hand sorting of contaminants, PET items are ground into flakes by a wet granulator. These ground flakes are transported to a silo for hot washing, which removes the last dirt and glue. In a final step at the washing station, these clean flakes are dried [49, 50].

The dry flakes are metered to an extrusion section, which melts, vacuum degasses, and finally grinds the clean washed flakes to produce food-grade natural PET pellets. These are stored in big sacks and are ready to be used to make new products [49, 50].

As for the MRF process, costs of the PET process are split into two categories, investments and operating costs, summarized over the factory's lifetime and then divided by the tons processed over the factory's lifetime to yield the cost per ton of plastic waste. The assumptions for the economic analysis are presented in Table 8.17 as well as Figure 8.30 in Section 8.3.

Table 8.17 Assumptions for Economic Analysis of Plastic Reprocessing Facility

Lifetime [years]	10
Yearly working hours [h]	6,240
Yearly PET capacity [t]	15,000
Total waste capacity (10 years) [t]	150,000
Separation efficiency [%]	91
Dollar to British Pound conversion rate [$/£]	1.46

Investment costs include building and site, machine, and additional equipment costs. Since the source of some of the investment costs were the United Kingdom, a Dollar to Pound conversion rate of 1.46 $/£ (as of April 26, 2016) was used for the calculations. Including costs for design and project management, civil engineering, land, and site work, the building and site investments added up to $4,140,000. The metering bin, bale breaker, washing station, and PET extrusion section costs were $25,852,220. These costs included conveyors and installation costs. For auxiliary equipment, further costs of $150,000 were incurred. Summed up, the investment costs for the PET reprocessing plant were **$30,142,220**, as shown in Table 8.18 (and in more detail in Figure 8.31 in Section 8.3) [47, 49, 51].

Table 8.18 Total Investment Costs of Plastic Reprocessing Facility

Cost Factor	Cost [$]
Building and site investment costs [$]	4,140,000
Machine investment costs [$]	25,852,220
Additional equipment investment costs [$]	150,000
Total investment costs [$]	**30,142,220**

Operating and maintenance costs include personnel salaries, maintenance, machine operating, and rolling stock operating costs. Assuming the factory runs 52 weeks per year, 5 days per week in 3 shifts with an effective working time of 8 hours each shift, the yearly working hours are 6,240 hours and result in personnel salaries of $2,193,290 per year, which are mainly for sorters. Maintenance costs of the machines are assumed as 5% of the investment costs and therefore amount to $1,292,611. Operating costs are caused by energy and water consumption especially for washing and extruding, which consume a high amount of energy, and lead to operating costs of $1,054,603 per year. Costs for diesel for the rolling stock are $111,507 per year. These four categories result in total operating and maintenance costs of **$4,652,011** per year, so **$46,520,110** after 10 years, as presented in Table 8.19. Detailed calculations of operating and maintenance costs are provided in Figure 8.32 in Section 8.3 [19, 20, 21, 47, 49].

Table 8.19 Total Operating and Maintenance Costs of Plastic Reprocessing Facility

Cost Factor	Cost [$]
Personnel salaries per year [$]	2,193,290
Maintenance costs per year [$]	1,292,611
Machine operating costs per year [$]	1,054,603
Rolling stock operating costs per year [$]	111,507
Yearly O&M costs [$]	**4,652,011**
Overall O&M costs (10 years) [$]	**46,520,110**

Summarizing investment and overall operating and maintenance expenses, processing 150,000 tons of PET in 10 years costs **$76,662,330**, or **$511.08** per ton of PET, as it is presented in Table 8.20. Since only 14.16% of plastic waste is PET, the costs of plastic processing of 1 t of plastic waste are **$72.37**.

Table 8.20 Total Costs of Plastic Reprocessing Facility

Cost Factor	Cost [$]
Investments costs [$]	30,142,220
Operating and maintenance costs [$]	46,520,110
Total costs (10 years; 150,000 t) [$]	**76,662,330**
Costs per ton of PET [$/t]	**511.08**
Costs per ton of plastic waste (14.16% PET) [$/t]	**72.37**

8.1.4.3 Revenues from Selling Recycled Plastic

The revenues from the PET recycling process come from selling the PET pellets produced to companies that produce new PET products for different applications, such as carbonated drink bottles (blow molding), food containers (thermoforming), or fibers (spinning).

In March 2020, the market price of 1 kg of PET pellets in the United States was on average $1.28 [52].

Calculating the revenues of plastic recycling, it needs to be remarked that only 14.16% of the plastic waste is PET, so only this 14.16% of 1 t of plastic waste is recycled in a best-case scenario, which means 141.6 kg out of 1 t. Furthermore, the efficiencies of both optical sorting during material recovery (89.18%) and manual sorting during PET processing need to be considered [47, 49, 53].

With an average price of $1.01/kg for recycled PET, revenues from recycling 1 t of plastic waste are $146.94, as shown in Table 8.21.

Table 8.21 Revenues from Recycling per Ton of PET Plastic Waste

Amount of PET in 1 t of plastic waste [kg]	141.60
Separation efficiency of plastic recycling [%]	89.18
Separation efficiency of PET processing [%]	91.00
Price of recycled PET [$/kg]	1.28
Total revenues per ton of plastic waste [$]	**146.94**

Small amounts of the other plastics are recycled. Currently, these other plastics produce close to no revenue due to a lack of reprocessing procedures for them in the United States. They are normally landfilled and so their revenues are not considered in this analysis [53].

8.1.4.4 Profitability

Finally, to calculate the profitability, the total costs per ton of waste need to be subtracted from its revenues.

The costs of both MRF and PET processing add up to $157.52 per ton of plastic waste, and the revenues of recycling 1 t of plastic waste are $146.94. Thus, the absolute profit from recycling 1 t of plastic waste is **–$10.58** (see Table 8.22).

Table 8.22 Profit from Recycling per Ton of PET Plastic Waste

Revenues per ton of plastic waste [$/t]	146.94
Total costs per ton of plastic waste [$/t]	157.52
Profit per ton of PET plastic waste [$/t]	**–10.58**

Since the profit is negative, recycling of plastic waste is not absolutely profitable. For this to have an absolute profit, the break-even price of recycled PET would need to be **$1.37**/kg instead of $1.28/kg.

This profitability depends significantly on the consumers. If plastic waste is not disposed of in the recycling trash can, plastic waste will not end up in the MRF, and so it will not be reprocessed at all. In this analysis, it was assumed that all waste was disposed of properly.

8.1.4.5 Influence of Oil Price on Profitability of Plastics Recycling

Oil is the most important raw material for plastics. As already mentioned, 1 kg of plastic requires about 2 kg of crude oil (including processing and raw materials for plastics). For this reason, the oil price has a great influence on the plastic price and the profitability of the whole plastic recycling process. The lower the oil price, the lower the price of the recycled plastic, the less profitable the recycling process becomes. Furthermore, a lower oil price reduces the costs of producing new, or virgin, plastic material, which is another challenge for plastics recycling.

In 2015 and 2016, low global oil prices significantly reduced the profitability of plastics recycling in the United States. In Newark, for example, the value for 1 bale of recycled plastics decreased from $230 to $112. One of the consequences was that Infinitus Energy, which had just opened a $35 million recycling center in Montgomery, Alabama, in 2014, shut that facility down in October 2014 since it was incurring losses only [54].

Figure 8.4 shows the changing price of 1 t of regrind PET compared to 1 barrel (158.9873 L) of oil. Since the original prices of the plastics were in Euros, the rate of 1 Dollar per Euro was taken from X-Rates. For a better comparison of different years, prices were inflation-adjusted based on the prices of October 2008 (both PET and oil) [55, 56, 57, 58].

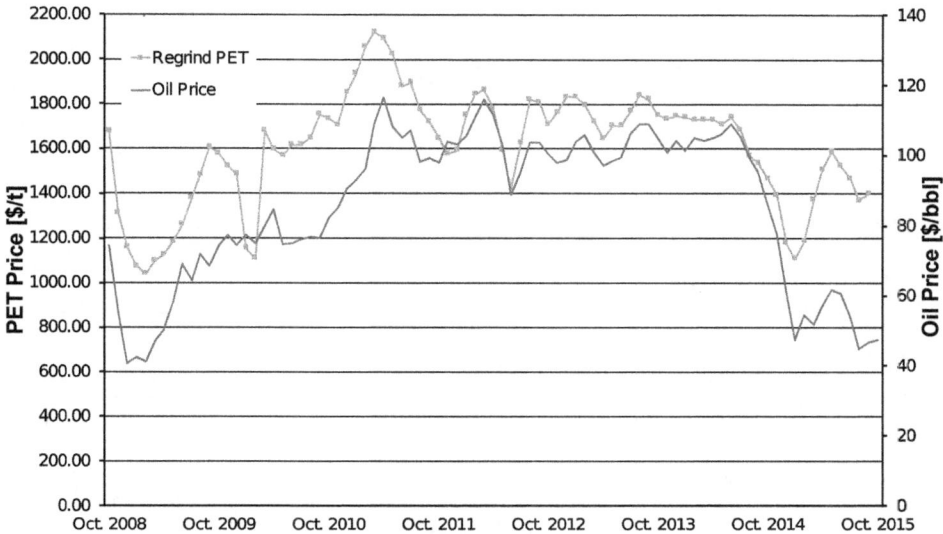

Figure 8.4 Oil (1 barrel (bbl) = 158.9873 L) and PET (1 t regrind) prices between October 2008 and October 2015 (inflation-adjusted)

Figure 8.4 demonstrates that the PET price is following the oil price changes, since the peaks of the regrind PET price graph line are lagging behind the peaks of the oil barrel price graph line by approximately 1 month. This can especially be observed in November 2009, October 2010, November 2014, and August 2015.

To analyze price correlations between PET and oil, a helpful tool is the calculation of the *Pearson product-moment correlation coefficient* $r_{x,y}$, which is a measure of degree of linear dependence between two variables. The calculation for $r_{x,y}$ is given in Equation 8.1.

$$r_{x,y} = \frac{\text{cov}(x,y)}{\sigma_x \sigma_y} = \frac{\sum\limits_{k=1}^{n} (x_k - \overline{x})(y_k - \overline{y})}{\sqrt{\left(\sum\limits_{k=1}^{n} (x_k - \overline{x})^2\right)\left(\sum\limits_{k=1}^{n} (y_k - \overline{y})^2\right)}} \tag{8.1}$$

The correlation coefficient ranges from $r_{x,y} \in [-1,1]$. A correlation coefficient $r_{x,y}$ of 1 implies a perfect relationship between x and y (a positive correlation). A value of -1 implies a perfect opposing relationship between the x and y (a negative correlation). A value of 0 implies no correlation between the variables [59] The correlation between the two prices is shown graphically in Figure 8.5.

For inflation-adjusted prices, the correlation coefficient is 0.84. This value reflects the results shown in Figure 8.4 and Figure 8.5: prices of PET and oil have a positive but not perfectly linear correlation.

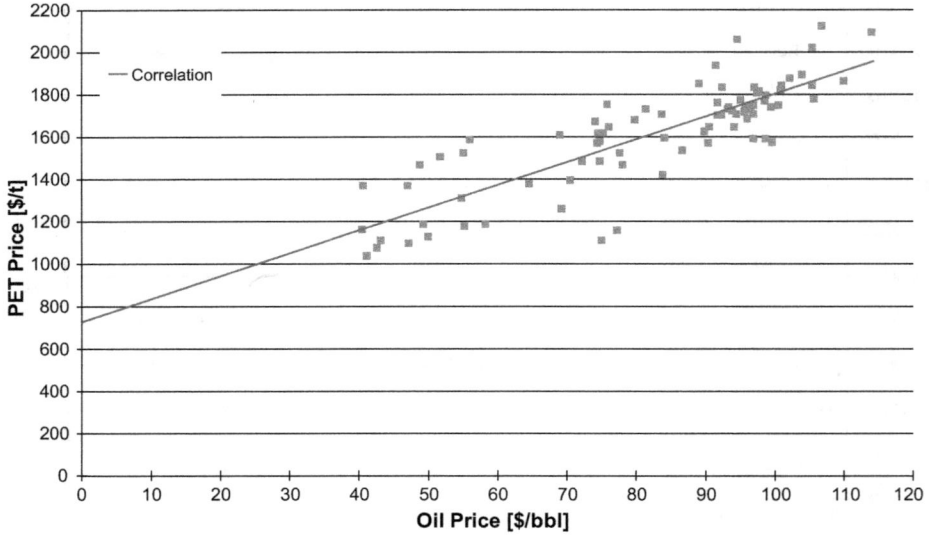

Figure 8.5 Correlation between PET price (per ton) and oil price (per barrel). bbl, barrel (158.9873 L)

Based on the correlation curve and knowing that the break-even price of recycled PET is $1.37 per kilogram, or $1,370.00 per ton, the required oil price for profitability is about $60.00 per barrel, as presented in Figure 8.6. In March 2020, the oil price was around $45 per barrel, which is too low to make plastic recycling profitable.

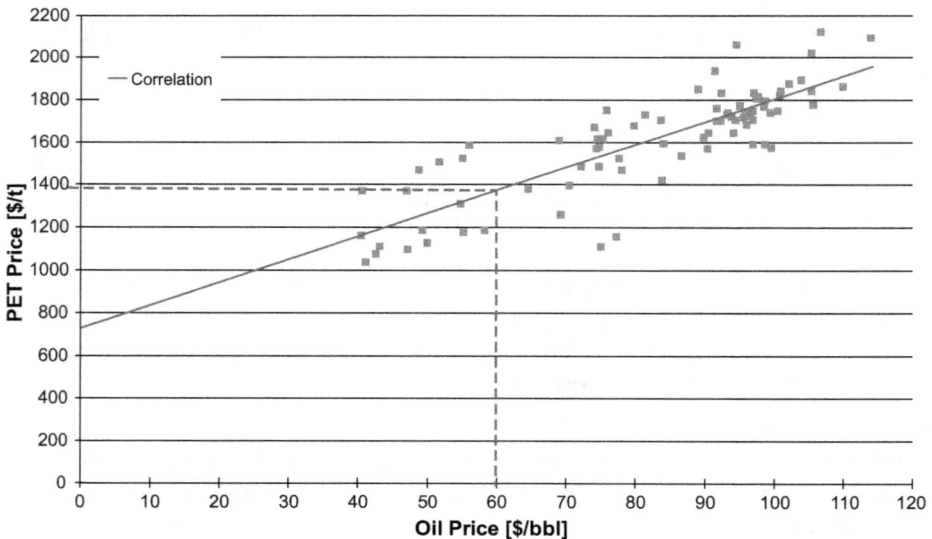

Figure 8.6 Required oil price per barrel for break-even recycled PET price per ton

8.1.5 Influence of China's Import Ban on Profitability of Plastics Recycling

The recycled plastics are only partially reused in the U.S. By 2017, more than 60% of recycled plastic waste was *exported*, as shown in Figure 8.7 [61].

Between 2017 and 2018 a significant shift in exported waste between countries can be observed. The import ban on plastic waste in China has reduced U.S. exports to China by almost 90%. Despite the increase in exports to other countries, especially Malaysia, Thailand, and Vietnam, approximately one third less plastic waste was exported in the first half of 2018 compared to the first half of 2017. This poses a major challenge for the U.S., as there were approximately 600,000 tons of plastic waste "left over" in 2018. For the recycling industry, a significant reduction in the selling price could be recorded: Between June 2018 and June 2019, the price for 1 kg of recycled PET pellets decreased from $1.59 to $1.37. This further reduces the economic viability of recycling in the U.S. Accordingly, new recycling options for plastic waste must be developed or plastic waste generally avoided [62].

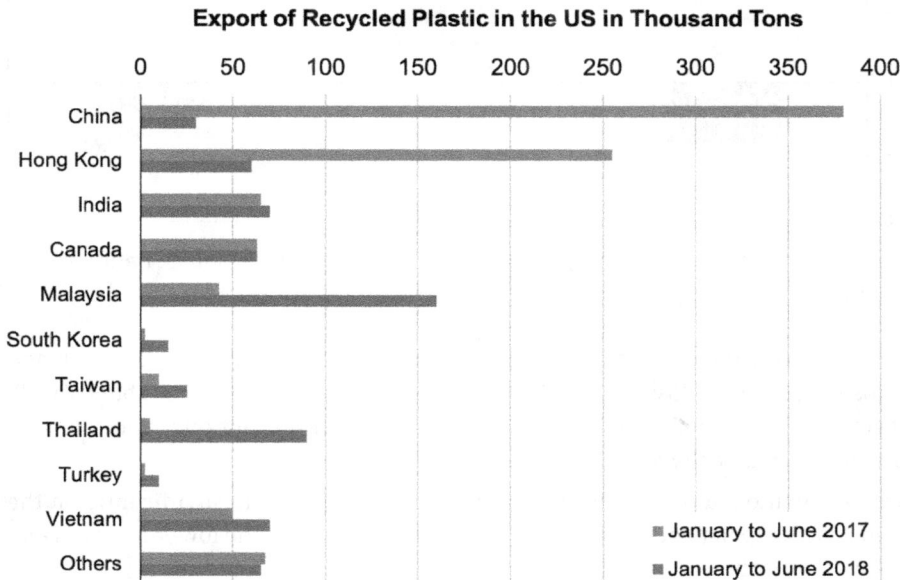

Figure 8.7 Distribution of exports of U.S. plastic waste by country

8.1.6 Conclusion: Economical Feasibility of Plastics Recycling

In the economic analysis presented using the static profit comparison method, the profit of handling plastic waste was analyzed for three different procedures. The profits of the three waste handling procedures for 1 t of waste are summarized in Figure 8.8. These numbers presume an ideal handling of plastic waste, which means all waste is disposed of properly, so that the plastic waste composition conforms with the plastic waste composition found in the MSW in 2017.

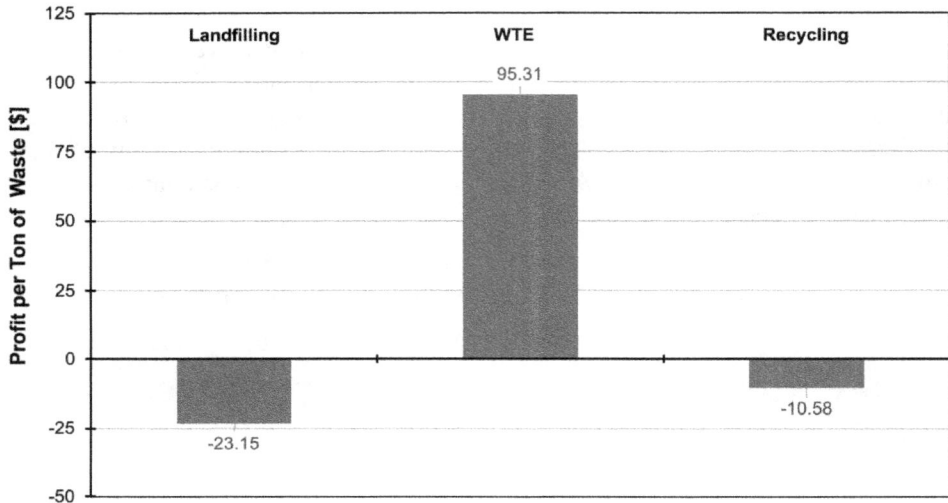

Figure 8.8 Profitability of recycling 1 t of plastic for three different waste handling procedures

Incineration with energy recovery (waste-to-energy) is the only handling method with absolute profitability (profit > $0.00) and therefore relative profitable as well. However, this is only valid in the case of incineration of plastics only and no other materials, which is really a theoretical assumption.

As already mentioned during this analysis, the numbers depend significantly on the size and yearly capacity of the facilities. The bigger the facility, the lower the costs and the higher the profit.

In the future, costs for landfilling will probably increase due to a lack of land available, so its profitability will further decrease.

The profitability of plastic recycling depends on two factors, which cannot be influenced by the factory: (1) the oil price and (2) the plastic recycling ratio of consumers, which depends on whether consumers recycle their waste properly, so that all waste plastic gets to the recovery factories to be recycled.

References for Section 8.1

[1] BWL Institut Basel. *Investitionsrechnung*. Basel, Switzerland. 2013.

[2] Shim, J. K., and Siegel, J. G. *Budgeting Basics and Beyond,* 3rd ed. John Wiley & Sons, Hoboken, NJ. 2009.

[3] Götze, U., Northcott, D., and Schuster, P. *Investment Appraisal: Methods and Models,* 2nd ed. Springer-Verlag, Berlin, Germany. 2015.

[4] Enseling, A. *Leitfaden zur Beurteilung der Wirtschaftlichkeit von Energiesparinvestitionen im Gebäudebestand.* Access Date: 2016/02/23. Available: *http://www.iwu.de/fileadmin/user_upload/dateien/energie/klima_altbau/leitfaden_wirtschaftlichkeit.pdf.*

[5] Bradu, M. Statistical-financial valuation methods of the investment projects. *Theoretical and Applied Economics Review.* vol. 508, pp. 49–52, 2007.

[6] Hering, E. *Investitions- und Wirtschaftlichkeitsrechnung für Ingenieure.* Springer Fachmedien Wiesbaden, Wiesbaden, Germany. 2014.

[7] Leeman, R. *Methoden der Wirtschaftlichkeitsanalyse von Energiesystemen.* Bundesamt für Konjunkturfragen (BfK), Impulsprogramme RAVEL, Bern, Switzerland. 1992.

[8] United States Senate. *Solid Waste Disposal Act (RCRA).* Washington, D. C. 2002.

[9] United States Environmental Protection Agency (EPA). *Regulatory impact analysis for the proposed revisions to the emission guidelines for existing sources and supplemental proposed new source performance standards in the municipal solid waste landfills sector.* Washington, D. C. 2015.

[10] United States Environmental Protection Agency (EPA). *Municipal Solid Waste Landfills—Economic Impact Analysis for the Proposed New Subpart to the New Source Performance Standards.* Washington, D. C. 2014.

[11] Duffy, D. P. Landfill Economics, Part II: Getting down to business. *MSW Management.* July/August 2005.

[12] Duffy, D. P. Landfill Economics, Part III: Closing up shop. *MSW Management.* September/October 2005.

[13] Duffy, D. P. Landfill Economics, Part I: Siting. *MSW Management.* vol. 15, issue 3, p. 118, May/June 2005.

[14] Caterpillar. *Specalog for 950H/962H—Wheel Loaders.* 2016.

[15] Caterpillar. *Specalog for 826G Series II—Landfill Compactor.* 2016.

[16] Caterpillar. *Specalog for D7R—Track-Type Tractor.* 2016.

[17] Eilrich, F. E., Doeksen G. A., and Van Fleet, H. *An Economic Analysis of Landfill Costs to Demonstrate the Economies of Size and Determine the Feasibility of a Community Owned Landfill in Rural Oklahoma.* Paper presented at the Southern Agricultural Economics Association, Mobile, AL, February 1–5, 2003.

[18] Glassdoor. *Waste Management Scale Attendant Salaries.* Access Date: 2016/04/13. Available: *https://www.glassdoor.com/Salary/Waste-Management-SCALE-ATTENDANT-Salaries-E2094_D_KO17,32.htm.*

[19] Salary.com. *Facility Manager Salaries.* Access Date: 2016/04/13. Available: *http://www1.salary.com/Facilities-Manager-Salary.html.*

[20] Salary.com. *Heavy Equipment Operator Salaries.* Access Date: 2016/04/13. Available: *http://www1.salary.com/Heavy-Equipment-Operator-Salaries.html.*

[21] Salary.com. *General Laborer Salaries.* Access Date: 2016/04/13. Available: *http://www1.salary.com/General-Laborer-Salary.html.*

[22] National Solid Wastes Management Association. *Modern Landfills: A Far Cry from the Past.* Washington, D. C. 2008.

[23] Gang, D. W. USA Today. *5 years after coal-ash spill, little has changed.* Access Date: 2016/04/13. Available: *http://www.usatoday.com/story/news/nation/2013/12/22/coal-ash-spill/4143995/.*

[24] Roland Berger Strategy Consultants. Trend 3: scarcity of resources. In: *Roland Berger Trend Compendium 2030.* 2015.

[25] National Solid Wastes Management Association. *Municipal Solid Waste Landfill Facts*. Washington, D. C. 2012.

[26] Michaels, T. *The 2014 ERC Directory of Waste-to-Energy Facilities*. Energy Recovery Council, Arlington, VA. 2014.

[27] Cekirge, H. M., Ouda, O. K. M., and Elhassan, A. Economic analysis of solid waste treatment plants using pyrolysis. *American Journal of Energy Engineering.* vol. 3, pp. 11–15, 2015.

[28] ICF International. *Economic Analysis of New Waste-to-Energy Facility in Metro Vancouver*. Paper prepared for Belkorp Environmental Services, Inc. 2014.

[29] Schneider, D. R., Lončar, D., and Bogdan, Ž. Cost analysis of waste-to-energy plant. *Strojarstvo.* vol. 52, pp. 369 –378, 2010.

[30] USforex. *Yearly Average Rates*. Access Date: 2016/04/13. Available: *http://www.usforex.com/forex-tools/historical-rate-tools/yearly-average-rates*.

[31] The World Bank. *Price level ratio of PPP conversion factor (GDP) to market exchange rate*. Access Date: 2016/04/13. Available: *http://data.worldbank.org/indicator/PA.NUS.PPPC.RF*.

[32] Glassdoor. *Maintenance Engineer Salaries*. Access Date: 2016/04/13. Available: *https://www.glassdoor.com/Salaries/maintenance-engineer-salary-SRCH_KO0,20.htm*.

[33] Salary.com. *Energy Manager Salaries*. Access Date: 2016/04/13. Available: *http://www1.salary.com/Energy-Manager-Salaries.html*.

[34] Salary.com. *Environmental Engineer III Salaries*. Access Date: 2016/04/13. Available: *http://www1.salary.com/Environmental-Engineer-III-Salaries.html*.

[35] U.S. Energy Information Administration. *Average Price of Electricity to Ultimate Customers by End-Use Sector.* Access Date: 2020/02/15. Available: *https://www.eia.gov/electricity/monthly/epm_table_grapher.php?t=epmt_5_6_a*.

[36] Themelis, N. J., and Mussche, C. *2014 Energy and Economic Value of Municipal Solid Waste (MSW), Including Non-Recycled Plastics (NRP) Currently Landfilled in the Fifty States*. Earth Engineering Center, Columbia University, New York, NY. 2014. pp. 1–40.

[37] Energy Information Administration, Office of Coal, Nuclear, Electric and Alternate Fuels. *Methodology for allocating municipal solid waste to biogenic and non-biogenic energy*. U.S. Department of Energy, Washington, D. C. 2007.

[38] Rand, T., Haukohl, J., and Marxen, U. *Municipal Solid Waste Incineration: A Decision Makers' Guide*. The World Bank, Washington, D. C. 2000.

[39] Davis, J., Haase, S., and Warren, A. *Waste-to-Energy Evaluation: U.S. Virgin Islands*. National Renewable Energy Laboratory (NREL), Golden, Colorado. 2011.

[40] Arsova, L., Van Haaren, R., Goldstein, N., Kaufman, S. M., and Themelis, N. J. The state of garbage in America. *BioCycle.* vol. 49, p. 22, 2008.

[41] Van Haaren, R., Themelis, N., and Goldstein, N. The state of garbage in America. *BioCycle.* vol. 51, pp. 16–23, 2010.

[42] Psomopoulos, C. S., Bourka, A., and Themelis, N. J. Waste-to-energy: a review of the status and benefits in USA. *Waste Management.* vol. 29, pp. 1718–1724, 2009.

[43] University of Cambridge and University of Cambridge-MIT Institute. *The ImpEE Project: Recycling of Plastics*. Department of Engineering, University of Cambridge, Cambridge, UK. 2005.

[44] Kessler Consulting Inc. *MRFing Our Way to Diversion: Capturing the Commercial Waste Stream*. Tampa, FL. 2009.

[45] Pellitteri Waste Systems. *Recycling Center Video – Watch your Recycling get sorted*. Access Date: 2016/04/13. Available: *http://www.pellitteri.com/news.jsp?id=1716*.

[46] Burns & McDonnell. *Mixed waste processing economic and policy study*. Austin, TX. Prepared for American Forest & Paper Association. 2015.

[47] Pressley, P. N., Levis, J. W., Damgaard, A., Barlaz, M. A., and DeCarolis, J. F. Analysis of material recovery facilities for use in life-cycle assessment. *Waste Management.* vol. 35, pp. 307–317, 2015.

[48] Gershman, Brickner & Bratton, Inc. *Materials Recovery Facility (MRF) Feasibility Report*. Fairfax, VA. Prepared for City of Tucson Environmental Services. 2008.

[49] Axion Consulting. *A Financial Assessment of Recycling Mixed Plastics in the UK: Financial modelling and assessment of mixed plastic separation and reprocessing (WRAP project MDP021)*. Prepared for Waste Recovery Action Programme (WRAP). 2009.

[50] Al-Salem, S., Lettieri, P., and Baeyens, J. Recycling and recovery routes of plastic solid waste (PSW): a review. *Waste Management*. vol. 29, pp. 2625–2643, 2009.

[51] X-Rates. *Currency-calculator: British Pound to Dollar*. Access Date: 2016/04/26. Available: *http://www.x-rates.com/calculator/?from=GBP&to=USD&amount=1*.

[52] Plasticnews. *Plastic Resin Pricing – Recycled Plastics*. Access Date: 2020/03/15. Available: *https://www.plasticsnews.com/resin/currentPricing/recycled-plastics*.

[53] United States Environmental Protection Agency (EPA). *Advancing Sustainable Materials Management: Facts and Figures 2013. Assessing Trends in Material Generation, Recycling and Disposal in the United States*. Washington, D. C. 2015.

[54] Gelles, D. *Losing a Profit Motive*. New York Edition: New York Times. February 12, 2016. p. B5. Available: *https://www.nytimes.com/2016/02/13/business/energy-environment/skid-in-oil-prices-pulls-the-recycling-industry-down-with-it.html?_r=0*.

[55] EUROINVESTOR. *Oil price development*. Access Date: 2016/04/13. Available: *http://www.euroinves tor.com/exchanges/gtis-energy/brent-oil/2327059*.

[56] Plasticker. *The Home of Plastics: Raw Materials & Prices*. Access Date: 2016/04/13. Available: *http://plasticker.de/preise/preise_monat_multi_en.php*.

[57] X-Rates. *US Dollar per 1 Euro Monthly average*. Access Date: 2016/04/13. Available: *http://www.x-rates.com/average/?from=EUR&to=USD&amount=1&year=2008*.

[58] US Inflation Calculator. *Historical Inflation Rates: 1914–2016*. Access Date: 2016/04/13. Available: *http://www.usinflationcalculator.com/inflation/historical-inflation-rates/*.

[59] Runkler, T. A. *Korrelation*. In: Data Mining: Methoden und Algorithmen intelligenter Datenanalyse. Vieweg+Teubner, Wiesbaden, Germany. 2010. pp. 55–63.

[60] Karidis, A. *Supply and Demands Drives Rising Tip Fees*. Access Date: 2020/03/15. Available: *https://www.waste360.com/landfill-operations/supply-and-demand-drives-rising-tip-fees*.

[61] Clarke, J. S., and Howard, E. *US Plastic Waste Exports to Developing Countries, Causing Environmental Problems at Home and Abroad*. Access Date: 2020/03/15. Available: *https://unearthed.green peace.org/2018/10/05/plastic-waste-china-ban-united-states-america/*.

[62] Dell, J. *157,000 Shipping Containers of U.S. Plastic Waste Exported to Countries with Poor Waste Management in 2018*. Access Date: 20/03/15. Available: *https://www.plasticpollutioncoalition.org/blog/2019/3/6/157000-shipping-containers-of-us-plastic-waste-exported-to-countries-with-poor-waste-management-in-2018*.

8.2 Optimization of Plastics Recycling

Recycling is the best option for handling plastic waste from an environmental point of view and can significantly contribute to minimizing air, soil, and marine pollution.

There are two central issues with recycling: on the one hand, only 9% of plastic waste in the United States is recycled at the moment due to technical limitations and, on the other hand, recycling is currently unprofitable from an economic point of view due to low oil prices. Recycling and selling 1 t of recycled plastic results in a loss of more than $10.

To improve both profitability and recycling rate, two process optimization possibilities are presented in this section.

8.2.1 Optimization I: Reduction of Sorting Processes

The first process optimization proposed is reducing the number of sorting processes. Therefore, the so-called *dual-stream recycling* would need to be implemented. Dual-stream recycling means that the plastic waste is directly separated by consumers in their households, which is similar to systems established in Europe. Consequently, the sorting process in the materials recovery facility (MRF) is not required anymore. The optimized process is shown in Figure 8.9 [1].

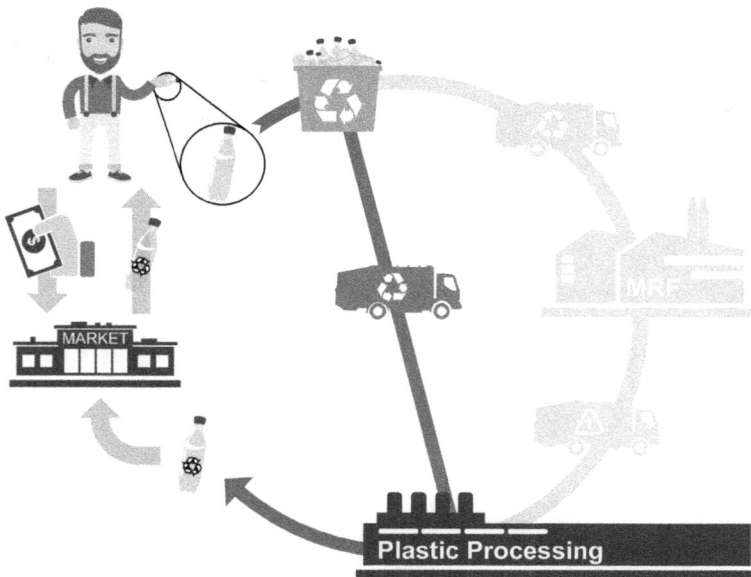

Figure 8.9 Optimization I: Dual-stream recycling

To calculate the profitability of the optimized process, the original profitability calculation of the plastic recycling process is used as a basis. The costs of polyethylene
terephthalate (PET) processing as well as the revenues realized by selling recycled
PET remain unchanged. Processing 1 t of plastic waste costs **$72.37** and the revenues
for sale of 1 t of recycled plastic are **$146.94**. But to handle plastic in the same facility,
additional machines and processes need to be installed. The additional costs are split
up in two main categories: investment costs (1) and operation and maintenance costs
(2). The assumptions for this optimization are shown in Table 8.23 and in more detail
in Figure 8.33 in Section 8.3.

Table 8.23 Optimization I: Assumptions

Lifetime [years]	10
Yearly working hours [h]	6,240
Yearly plastic waste handling [t]	100,000
Total plastic waste capacity (10 years) [t]	1,000,000
Yearly PET capacity [t]	15,000
Total PET waste capacity (10 years) [t]	150,000
Separation efficiency [%]	91

Additional investment costs are split up in building and site, machine, and equipment
costs. To handle plastic waste in only one facility, additional land, site work, and
buildings as well as a scale house are required. These building and site costs amount
to $1,775,000. Furthermore, three new machines need to be installed: a metering bin,
an optical PET sorting machine, and a baler. The investment costs of all machines add
up to $925,000. For additional conveyors, rolling stock, and waste collection cars, total
costs are $1,250,000. As presented in Table 8.24, total additional investment costs are
$3,950,000 (see also Figure 8.34 in Section 8.3) [2, 3, 4].

Table 8.24 Optimization I: Additional Investment Costs

Cost Factor	Cost [$]
Additional building and site investment costs [$]	1,775,000
Additional machine investment costs [$]	925,000
Additional equipment investment costs [$]	1,250,000
Total additional investment costs [$]	**3,950,000**

Additional operating and maintenance costs are salaries of the additional personnel, operating and maintenance costs of the machines and the rolling stock, and especially transportation and collection costs. Yearly operating and maintenance costs are **$5,713,797**, so overall **$57,137,976**, as presented in Table 8.25 and in more detail in Figure 8.35 in Section 8.3 [3, 5, 6, 7, 8].

Table 8.25 Optimization I: Additional Operating and Maintenance (O&M) Costs

Cost Factor	Cost [$]
Personnel salaries per year [$]	963,000
Facility costs per year [$]	250,000
Machine O&M costs per year [$]	68,417
Rolling stock O&M costs per year [$]	748,380
Transportation and collection costs [$]	3,684,000
Yearly O&M costs [$]	**5,713,797**
Overall O&M costs (10 years) [$]	**57,137,976**

Summarizing both additional investment and operating and maintenance costs, total additional costs are **$61,087,976**. Since 100,000 t of plastic waste must be handled per year in this new facility area (to gain 15,000 t of PET waste, around 100,000 t of plastic waste has to be sorted), the additional costs of 1 t of plastic waste are **$61.09**.

Knowing that the revenues of recycling 1 t of plastic waste are **$146.94** and the costs for further processing the plastic waste are **$72.37**, the profitability of this optimization is calculated in Table 8.26.

Table 8.26 Total Profit per Ton of Plastics Recycled

Revenues per ton of plastics recycled [$/t]	146.94
Sorting [$]	61.09
PET processing [$]	72.37
Profit per ton of plastics recycled [$/t]	**13.48**

Table 8.26 shows that the profit of this optimization would be **$13.48** per ton of plastic waste recycled, so this process optimization has absolute profitability. Compared to the current plastic recycling procedure, the profit of the optimization is **$24.06** higher.

But this optimization has two risks, which cannot be included in the calculations. First, this optimized process depends even more on the **disposal behavior of consumers** than the current recycling process already does. If consumers not only have to distinguish between recyclables and non-recyclables but also between plastics, paper, glass, and metals, the risk of not properly disposing of plastic waste increases.

Second, the optimization has **hidden costs** for the transportation and collection of waste. The analysis only considered the recycling of plastic waste. In daily life, paper, glass, and metals should be recycled as well. If all of these materials were collected individually, transportation and collection costs would increase immensely.

8.2.2 Optimization II: Upcycling of Plastic Waste by Blending PP and LDPE

The second optimization possibility considered in this book includes the *extension of the analyzed PET reprocessing* to recycling polypropylene (PP) and low-density polyethylene (LDPE) by blending them into a compound. The process is presented in Figure 8.10.

Due to its desirable physical properties such as high tensile strength, high stiffness, and high chemical resistance, PP has been widely used as a packaging material. However, it shows poor impact strength at low temperatures and is susceptible to environmental *stress cracking*. LDPE is mostly used for bags and packaging films. Owing to its low mechanical properties but ease of processing, it is recycled and used for garbage bags. Therefore, blends of LDPE and PP have become a subject of great economic interest to improve the processing and mechanical properties of PP [9].

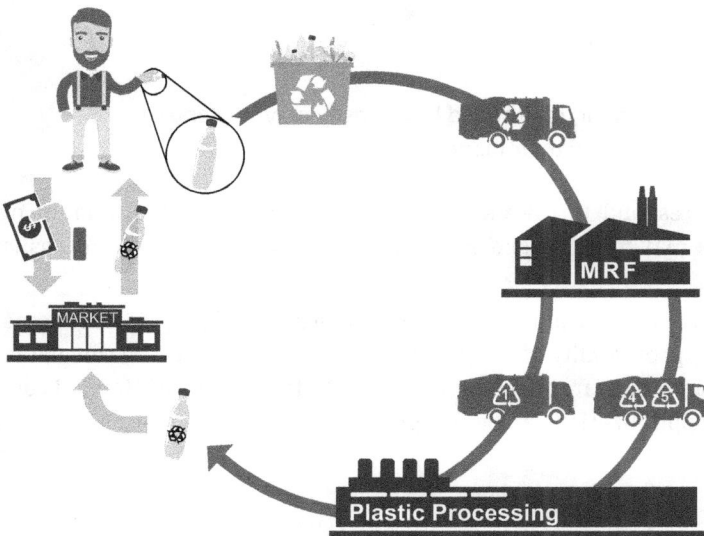

Figure 8.10 Optimization II: Extension of the process by blending recycled polypropylene (PP) and low-density polyethylene (LDPE)

Combinations of LDPE and PP are frequently found in polymer waste streams. But since their densities are very similar (PP: 946 kg/m³, LDPE: 940 kg/m³), they cannot be easily separated from each other by conventional sorting methods. Another motivation for blending recycled PP–LDPE is the high impact it has. PP and LDPE accounted for almost 50% of the total plastic waste in 2017. Not recycling these two plastics means wastage of important resources. However, it needs to be determined if blending recycled PP and LDPE would be economically profitable and as a result would optimize the plastic recycling process [10, 11, 12].

Material tests showed that recycled PP and LDPE blends have material properties similar to virgin PP–LDPE blends. The effects of processing on the properties of recycled PP, recycled LDPE, and their blends were investigated using *melt flow index (MFI) measurements*. These MFI measurements showed an increase with increasing amounts of rLDPE in blends, thus a decrease of the viscosity with increasing amounts of rLDPE, which is due mainly to the higher degradation of rLDPE, as shown in Figure 8.11.

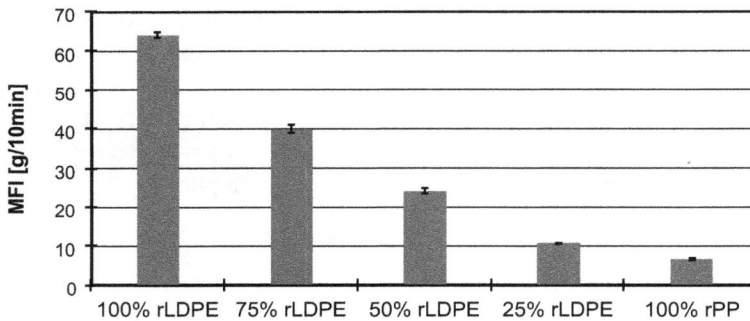

Figure 8.11 Melt flow index (melt) of recycled low-density polyethylene (rLDPE), recycled polypropylene (rPET), and their blends

These results also correspond to the mechanical tests. The higher the amount of rLDPE, the higher the maximum tensile strain but the lower the maximum tensile stress (Figure 8.12).

Accordingly, blending recycled PP and LDPE can improve material properties to some extent. To calculate the profitability of optimization II, additional costs are calculated first. Afterwards, the revenues from selling rLDPE–rPP blends are calculated. From these, the profit of the optimization can be computed.

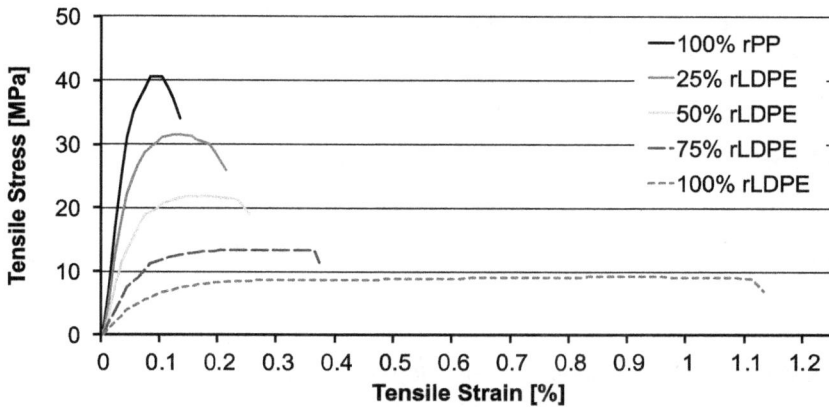

Figure 8.12 Tensile stress and tensile strain of recycled low-density polyethylene (rLDPE), recycled polypropylene (rPP), and their blends

8.2.2.1 Additional Costs of LDPE–PP Recycling

To calculate the costs of optimization II, the plastic recycling process in Section 8.1.4 serves as a basis. The important assumptions are presented in Table 8.13, Table 8.17, and Table 8.27.

Table 8.27 Optimization II: Assumptions

Lifetime [years]	10
Percentage of PET in plastic waste [%]	14.39
Percentage of LDPE–PP in plastic waste [%]	45.70
Average price of recycled PET pellets [$/kg]	1.26
Average price of recycled LDPE–PP pellets [$/kg]	0.66
Total waste capacity (10 years) [t]	150,000
Separation efficiency [%]	91

First, the MRF process is extended by machines that are required to sort LDPE and PP. Since their densities are nearly the same and the optical sorter uses material densities to distinguish between different plastics, only one new sorter is required. Total investment as well as operating and maintenance costs are **$1,170,892**, or **$0.98** per t of plastic, as shown in Table 8.28 and in more detail in Figure 8.36 in Section 8.3 [2].

Table 8.28 Optimization II: Additional Costs of Low-Density Polyethylene and Poly-propylene (LDPE–PP) Blending for Materials Recovery Facility (MRF)

Cost Factor	Cost [$]
Additional investment costs [$]	900,000
Additional O&M costs [$]	265,734
Overall additional MRF costs [$]	**1,170,892**
Additional MRF costs per ton [$/t]	**0.98**

Second, the plastic waste process needs to be adapted. Since plastics are sorted in the MRF first and transported separately to the plastic processing factory, PET and LDPE–PP are processed independently from each other. That means that a second metering bin, bale breaker, and washing station are required, as well as an LDPE–PP extruder. The yearly capacity for LDPE–PP is the same as for PET. Thus, investment costs for the metering bin, bale breaker, and washing station are the same as for PET. The LDPE–PP extruder is a bit cheaper than the PET extruder. Total machine investment costs of plastic processing are $24,069,560 (conveyers and installation included). In addition, facility space and rolling stock need to be expanded. These investments result in total costs of **$26,069,560** (Table 8.29 and Figure 8.37 in Section 8.3) [2, 13].

Table 8.29 Optimization II: Additional Investment Costs for Plastic Processing

Cost Factor	Cost [$]
Building and site investment costs [$]	1,950,000
Machine investments costs [$]	24,069,560
Additional equipment investment costs [$]	50,000
Total investment costs [$]	**26,069,560**

Since LDPE–PP requires an individual process, operation and maintenance costs are also similar to PET processing. All materials are processed in the same facility; therefore, no additional plant manager, marketing manager, and maintenance engineer is required. Another big difference between PET and LDPE–PP processing is electricity consumption of the extrusion process: LDPE–PP has a lower melting temperature than PET and consequently a lower electricity consumption. Yearly operation and maintenance costs of LDPE–PP processing are **$3,792,563**, thus **$37,925,630** after 10 years (Table 8.30 and Figure 8.38 in Section 8.3) [2, 13].

Table 8.30 Optimization II: Additional Operating and Maintenance (O&M) Costs of Plastic Processing

Cost Factor	Cost [$]
Personnel salaries per year [$]	1,500,000
Machine maintenance costs per year [$]	1,203,478
Machine operating costs per year [$]	977,578
Rolling stock operating costs per year [$]	111,507
Yearly additional O&M costs [$]	**3,792,563**
Overall additional O&M costs (10 years) [$]	**37,925,630**

Summarizing all additional investment and operating and maintenance costs over the factory's lifetime, total additional costs of the optimization process are $63,995,190, or $426.63 per t of LDPE–PP waste. Assuming a combined LDPE–PP content of 45.70% in the plastic waste, additional processing costs per 1 t of LDPE–PP plastic are $193.95 (Table 8.31).

Table 8.31 Optimization II: Additional Costs per Ton for Plastic Processing

Cost Factor	Cost [$]
Investments costs [$]	26,069,560
Operating and maintenance costs [$]	37,925,630
Total costs (10 years – 150,000 t) [$]	**63,995,190**
Costs per ton of LDPE–PP [$/t]	**426.63**
Costs per ton of plastic waste (45.46%) [$/t]	**193.95**

8.2.2.2 Additional Revenues of LDPE–PP Recycling

To calculate the revenues of the LDPE–PP recycling process, the main task is to determine the selling price of rLDPE–rPP blends.

The results of the material analysis agreed with the behavior of virgin material: a higher amount of PP in blends leads to increased stiffness and strength of the material, which are desirable physical properties for packaging material. However, a greater portion of LDPE in the blends increased the maximum strain of material. Depending on the choice of application for the material, the ability of the material to elongate can also be an important criterion.

This shows that the value of a product is regulated by its demand. Determining an explicit selling price based on material properties is, also due to a lack of experience in this area, relatively arbitrary.

Since the ratio between LPDE (22.84% of plastic waste) and PP (22.62% of plastic waste) in the plastic waste is nearly identical, a ratio of 50% LDPE and 50% PP in the blend is assumed. For this reason, the price of recycled LDPE–PP pellets used for the following calculations is $0.66 per kg, which is the mean value of rLDPE ($0.52 per kg) and rPP ($0.80 per kg) for March 2020 [14].

Calculating the revenues of plastic recycling, it needs to be remarked again that only 455 kg of 1 t of plastic waste is LDPE–PP. The efficiencies of both the optical sorting at the MRF and the manual sorting during LDPE–PP processing need to be included as well. With an average price of $0.66 per kg, additional revenues through Optimization II are **$236.53** (Table 8.32).

Table 8.32 Optimization II: Additional Revenues

Amount of LDPE–PP in plastic waste [kg]	454.60
Separation efficiency of MRF process [%]	86.45
Separation efficiency of PET processing [%]	91.00
Price of recycled LDPE–PP [$/kg]	0.66
Total revenues from recycling LDPE–PP [$]	**236.53**

8.2.2.3 Total Profit of Optimization II

Totaling revenues and costs from both the PET process (Section 8.1.4) and optimization II and subtracting the costs from the revenues, the profit from recycling 1 t of plastic waste is **$31.65** (Table 8.33).

Table 8.33 Optimization II: Total Profit per Ton of Plastics Recycled

Revenues per ton of plastics [$/t]	383.47
Costs per ton of plastics [$/t]	351.82
Profit per ton of plastics recycled [$/t]	**31.65**

With optimization II, the profit would increase by **$42.50** per ton and make plastic recycling absolutely profitable. The implementation of an LDPE–PP process is relatively expensive, but since the additional costs of the MRF are very low, the profit increases significantly.

But as already mentioned earlier in this book, this is an ideal case with the prerequisite that all plastic is sorted and disposed of properly by consumers. Furthermore, the selling price of the rLDPE–rPP is of great importance for the revenue of the optimization process, and thus the profitability of the whole plastics recycling process. The **break-even selling price of rLDPE–rPP is $0.57/kg**. If the selling price for the blend were lower, the optimization would not be profitable. Furthermore, the price of oil is an important factor for optimization II as well. The dependence of rLDPE and rPP prices on the price of oil is the same as for rPET. Assuming the oil prices of April 2016, optimization II turns plastic recycling into a profitable process.

8.2.3 Optimization III: Increasing the Recycling Rate

As presented in Section 8.2.2, upcycling[1] of plastic waste by blending several polymers improves the recycling process from both an economic and ecological point of view. In an optimal scenario, the recycling rate of plastic waste in the United States could be increased by 45.46% based on the numbers of 2017.

However, the problem of this upcycling scenario is that the blended plastics become number 7 plastics (mixed plastic) and are therefore difficult to recycle a second or even a third time. Even if blending can improve the material properties, it is still desirable from an ecological perspective to recycle all plastics separately. Thus, multiple reprocessing cycles could be executed.

As mentioned in Section 8.2.2, separating PP (density: $946 \, kg/m^3$) and LDPE (density: $940 \, kg/m^3$) is nearly impossible by conventional and economically efficient sorting methods. However, research regarding the separation of each plastic type has intensified. One organization in the United States trying to increase the recycling rate of plastics is the *Closed Loop Fund (CLF)*. It is a social impact fund that invested $100 million to increase the recycling of products and packaging. It is their aim by 2025 to proof replicable recycling processes that will help unlock additional investments in recycling [15, 16].

One of their approaches is the improvement of recycling of numbers 3 to 7 plastics. Even in major markets, no viable options for recycling these plastics exist. As a result, much of this material is being landfilled. Therefore, CLF partnered with QRS Recycling and Canusa Hershman Recycling to create a state-of-the-art plastic recovery factory in Maryland with a capacity to handle more than 50,000 t/year. This $15 million project was supported by a $2 million investment from CLF and $13 million from other private sources [16]. In QRS Recycling's Maryland facility, mixed numbers 3 to 7 plastics in bales from single-stream MRFs are transformed into a high-quality, postconsumer PP and PE flake that can be used by various plastics manufacturers [16].

---

[1] Upcycling is the process of transforming waste materials into new materials or products of better quality or for better environmental value.

One of the most significant challenges of this business model besides the sorting process is contamination. Bales of mixed plastics (numbers 3 to 7) often contain a high level of contamination. Certain materials out of these residues can have a costly impact on the process [16].

Furthermore, the profitability of the business model is still unclear. Recycled LDPE and PP could probably be sold for $0.52 per kg (LDPE) and $0.80 per kg (PP), but the processing costs are difficult to calculate. In addition, for some feedstocks (mainly numbers 3 and 6 plastics), there is not yet any market and so value for the recycled plastics yet. Thus, these plastics would probably be landfilled, which creates negative impacts both environmentally and ecologically [16].

However, an increasing recycling rate of plastics has to be the aim of the plastics industry. The environmental value is the highest of all optimization scenarios, mainly because plastics can be recycled several times. And even if the process might not be profitable yet, the economic potential is high.

References for Section 8.2

[1] Kessler Consulting Inc. *MRFing Our Way to Diversion: Capturing the Commercial Waste Stream*. Tampa, FL. 2009.

[2] Pressley, P.N., Levis, J.W., Damgaard, A., Barlaz, M.A., and DeCarolis, J.F. Analysis of material recovery facilities for use in life-cycle assessment. *Waste Management*. vol. 35, pp. 307–317, 2015.

[3] Gershman; Brickner & Bratton Inc. *Materials Recovery Facility (MRF) Feasibility Report*. Fairfax, VA. 2008.

[4] Burns & McDonnell. *Mixed waste processing economic and policy study*. Austin, TX. Report prepared for the American Forest & Paper Association. 2015.

[5] Glassdoor. *Waste Management Scale Attendant Salaries*. Access Date: 2016/04/13. Available: *https://www.glassdoor.com/Salary/Waste-Management-SCALE-ATTENDANT-Salaries-E2094_D_KO17,32.htm*.

[6] Salary.com. *Facility Manager Salaries*. Access Date: 2016/04/13. Available: *http://www1.salary.com/Facilities-Manager-Salary.html*.

[7] Salary.com. *Heavy Equipment Operator Salaries*. Access Date: 2016/04/13. Available: *http://www1.salary.com/Heavy-Equipment-Operator-Salaries.html*.

[8] Salary.com. *General Laborer Salaries*. Access Date: 2016/04/13. Available: *http://www1.salary.com/General-Laborer-Salary.html*.

[9] Utracki, L.A. *Commercial polymer blends*: Springer Science & Business Media, New York, NY. 1998.

[10] Yin, S., Tuladhar, R., Shi, F., Shanks, R.A., Combe, M., and Collister, T. Mechanical reprocessing of polyolefin waste: a review. *Polymer Engineering & Science*. vol. 55, pp. 2899–2909, 2015.

[11] Salih, S.E., Hamood, A.F., and Alsabih, A.H. Comparison of the characteristics of LDPE: PP and HDPE: PP polymer blends. *Modern Applied Science*. vol. 7, p. 33, 2013.

[12] Mastalygina, E.E., Popov, A.A., Kolesnikova, N.N., and Karpova, S.G. *International Journal of Plastics Technology*. vol. 19, pp. 68–83, 2015.

[13] Axion Consulting. *Financial modelling and assessment of mixed plastic separation and reprocessing (WRAP project MDP021)*. Report prepared for Waste Resources Action Programme (WRAP). 2009.

[14] Plastics News. *Current Resin Pricing*. Access Date: 2020/03/21. Available: *https://www.plasticsnews.com/resin/currentPricing/recycled-plastics*.

[15] Closed Loop Fund. About the Closed Loop Fund. Access Date: 2020/05/06. Available: *https://www.closedlooppartners.com/about-us/*.

[16] Closed Loop Fund. *Impact Report January 2015—September 2016*. 2016.

8.3 Appendix to the Second Edition

This section shows the detailed numbers and calculations of the economic analysis in Section 8.1 (corresponding to Chapter 4 of the 2nd edition) and the optimization scenarios in Section 8.2 (corresponding to Chapter 6 of the 2nd edition).

Since these calculations are based and depend on several assumptions (e. g., efficiencies, size of factories, waste capacity, exchange rate, salaries), the tables enable the analysis of different scenarios for landfilling, incineration with energy recovery, and recycling of plastic waste.

> ☑ Yet more detailed spreadsheets used in these analyses can be downloaded from *plus.hanser-fachbuch.de/en,* by entering the code provided on the last page of this book. These can be adapted to allow you to obtain results for your own ideas and scenarios.

8.3.1 Economic Analysis of Landfilling

Buildings	Area [sq ft]	Costs [$/sq ft]			Total Costs [$]		
		Min.	Max.	Average	Min.	Max.	Average
Maintenance Buildings	10,000	50.00	70.00	60.00	500,000.00	700,000.00	600,000.00
Office Buildings	3,000	60.00	100.00	80.00	180,000.00	300,000.00	240,000.00
Shacks and Tool Sheds	1,000	10.00	20.00	15.00	10,000.00	20,000.00	15,000.00

IT	Pieces	Costs per Unit [$/piece]			Total Costs [$]		
		Min.	Max.	Average	Min.	Max.	Average
Associated Computer Systems	1	50,000.00	75,000.00	62,500.00	50,000.00	75,000.00	62,500.00
Modular Truck Scales	1	50,000.00	75,000.00	62,500.00	50,000.00	75,000.00	62,500.00

Roads	Area [sq ft]	Costs [$/sq ft]			Total Costs [$]		
		Min.	Max.	Average	Min.	Max.	Average
Perimeter Access Roads (Gravel)	120,000	1.00	2.00	1.50	120,000.00	240,000.00	180,000.00
Perimeter Access Roads (Asphalt)	0	6.00	9.00	7.50	0.00	0.00	0.00

Security Barrier	Length [ft]	Costs [$/.ft.]			Total Costs [$]		
		Min.	Max.	Average	Min.	Max.	Average
Fence	6,000	10.00	20.00	15.00	60,000.00	120,000.00	90,000.00
Gates	5	1,000.00	2,000.00	1,500.00	5,000.00	10,000.00	7,500.00
Signages (200 ft Intervals)	30 pc.	10.00	20.00	15.00	300.00	600.00	450.00

Washing	Pieces	Costs per Unit [$/piece]			Total Costs [$]		
		Min.	Max.	Average	Min.	Max.	Average
Wheel Washing Facilities	1	200,000.00	250,000.00	225,000.00	200,000.00	250,000.00	225,000.00

Liner Construction	Area [acre]	Costs [$/acre]			Total Costs [$]		
		Min.	Max.	Average	Min.	Max.	Average
Clear and Grub	33.50	1,000.00	3,000.00	2,000.00	33,500.00	100,500.00	67,000.00
Site Survey	33.50	5,000.00	8,000.00	6,500.00	167,500.00	268,000.00	217,750.00
Excavation	33.50	100,000.00	330,000.00	215,000.00	3,350,000.00	11,055,000.00	7,202,500.00
Perimeter Berm	33.50	10,000.00	16,000.00	13,000.00	335,000.00	536,000.00	435,500.00
Clay Liner	33.50	32,000.00	162,000.00	97,000.00	1,072,000.00	5,427,000.00	3,249,500.00
Geomembrane	33.50	24,000.00	35,000.00	29,500.00	804,000.00	1,172,500.00	988,250.00
Geocomposite	33.50	33,000.00	44,000.00	38,500.00	1,105,500.00	1,474,000.00	1,289,750.00
Granular Soil	33.50	48,000.00	64,000.00	56,000.00	1,608,000.00	2,144,000.00	1,876,000.00
Leachate System	33.50	8,000.00	102,000.00	55,000.00	268,000.00	3,417,000.00	1,842,500.00
QA/QC	33.50	75,000.00	100,000.00	87,500.00	2,512,500.00	3,350,000.00	2,931,250.00

Total Costs					12,431,300.00	30,734,600.00	21,582,950.00

Figure 8.13 Economic analysis of landfill: construction costs (imperial)

Buildings	Area [sq m]	Costs [$/sq m]			Total Costs [$]		
		Min.	Max.	Average	Min.	Max.	Average
Maintenance Buildings	929	538.18	753.46	645.82	500,000.00	700,000.00	600,000.00
Office Buildings	279	645.82	1,076.36	861.09	180,000.00	300,000.00	240,000.00
Shacks and Tool Sheds	93	107.64	215.27	161.45	10,000.00	20,000.00	15,000.00

IT	Pieces	Costs per Unit [$/piece]			Total Costs [$]		
		Min.	Max.	Average	Min.	Max.	Average
Associated Computer Systems	1	50,000.00	75,000.00	62,500.00	50,000.00	75,000.00	62,500.00
Modular Truck Scales	1	50,000.00	75,000.00	62,500.00	50,000.00	75,000.00	62,500.00

Roads	Area [sq m]	Costs [$/sq m]			Total Costs [$]		
		Min.	Max.	Average	Min.	Max.	Average
Perimeter Access Roads (Gravel)	11,149	10.76	21.53	16.15	120,000.00	240,000.00	180,000.00
Perimeter Access Roads (Asphalt)	0	64.58	96.87	80.73	0.00	0.00	0.00

Security Barrier	Length [m]	Costs [$/m]			Total Costs [$]		
		Min.	Max.	Average	Min.	Max.	Average
Fence	1,829	32.81	65.62	49.21	60,000.00	120,000.00	90,000.00
Gates	2	3,280.80	6,561.60	4,921.20	5,000.00	10,000.00	7,500.00
Signages (200 ft Intervals)	30 pc.	10.00	20.00	15.00	300.00	600.00	450.00

Washing	Pieces	Costs per Unit [$/piece]			Total Costs [$]		
		Min.	Max.	Average	Min.	Max.	Average
Wheel Washing Facilities	1	200,000.00	250,000.00	225,000.00	200,000.00	250,000.00	225,000.00

Liner Construction	Area [sq m]	Costs [$/sq m]			Total Costs [$]		
		Min.	Max.	Average	Min.	Max.	Average
Clear and Grub	13.56	2,471.05	7,413.16	4,942.11	33,500.00	100,500.00	67,000.00
Site Survey	13.56	12,355.27	19,768.43	16,061.85	167,500.00	268,000.00	217,750.00
Excavation	13.56	247,105.41	815,447.84	531,276.63	3,350,000.00	11,055,000.00	7,202,500.00
Perimeter Berm	13.56	24,710.54	39,536.87	32,123.70	335,000.00	536,000.00	435,500.00
Clay Liner	13.56	79,073.73	400,310.76	239,692.25	1,072,000.00	5,427,000.00	3,249,500.00
Geomembrane	13.56	59,305.30	86,486.89	72,896.10	804,000.00	1,172,500.00	988,250.00
Geocomposite	13.56	81,544.78	108,726.38	95,135.58	1,105,500.00	1,474,000.00	1,289,750.00
Granular Soil	13.56	118,610.60	158,147.46	138,379.03	1,608,000.00	2,144,000.00	1,876,000.00
Leachate System	13.56	19,768.43	252,047.52	135,907.97	268,000.00	3,417,000.00	1,842,500.00
QA/QC	13.56	185,329.06	247,105.41	216,217.23	2,512,500.00	3,350,000.00	2,931,250.00
Total Costs					**12,431,300.00**	**30,734,600.00**	**21,582,950.00**

Figure 8.14 Economic analysis of landfill: construction costs (metric)

Personnel	Units	Unit Costs [$/unit]	Total Costs [$]
Facility Manager	2	90,000.00	180,000.00
Equipment Operators	4	60,000.00	240,000.00
Scale House Attendant	2	40,000.00	80,000.00
General Laborers	4	30,000.00	120,000.00

Equipment Operating Costs	Usage [h]	Cost per Hour [$/h]	Total Costs [$]
Bulldozer	3,180	55.00	174,900.00
Compactor	3,180	50.00	159,000.00
Front-End-Loader	3,180	50.00	159,000.00
Grader	3,180	30.00	95,400.00

Site Repairs and Maintenance	Units	Unit Costs [$/unit]	Total Costs [$]
Perimeter Access Roads (Gravel)	1	160,000.00	160,000.00

Leachate Collection and Treatment	Gallons	Costs per Gallon [$/gallon]	Total Costs [$]
Leachate Collection and Treatment Costs	1,500,000.00	0.02	30,000.00

Others	Units	Unit Costs [$/unit]	Total Costs [$]
Environmental Sampling and Monitoring	1	60,000.00	60,000.00
Engineering Services (Consulting, in-house staff)	1	120,000.00	120,000.00

Operation Costs per Year [$]			**1,578,300.00**
Total Operation Costs (11 years) [$]			**17,361,300.00**

Figure 8.15 Economic analysis of landfill: operations costs

Task	Area [acre]	Costs [$/acre]			Total Costs [$]		
		Min.	Max.	Average	Min.	Max.	Average
Final Grade Survey	34.00	3,000.00	6,000.00	4,500.00	102,000.00	204,000.00	153,000.00
Gas Management Layer	34.00	24,000.00	32,000.00	28,000.00	816,000.00	1,088,000.00	952,000.00
Compacted Clay Cap	34.00	26,000.00	51,000.00	38,500.00	884,000.00	1,734,000.00	1,309,000.00
Geomembrane Cap	34.00	18,000.00	23,000.00	20,500.00	612,000.00	782,000.00	697,000.00
Geocomposite	34.00	33,000.00	44,000.00	38,500.00	1,122,000.00	1,496,000.00	1,309,000.00
Cover and Vegetative Soil	34.00	13,000.00	26,000.00	19,500.00	442,000.00	884,000.00	663,000.00
Seed, Mulch, and Fertilizer	34.00	1,000.00	2,000.00	1,500.00	34,000.00	68,000.00	51,000.00
Gas Management System	34.00	29,000.00	35,000.00	32,000.00	986,000.00	1,190,000.00	1,088,000.00
Run-Off Control System	34.00	5,000.00	7,000.00	6,000.00	170,000.00	238,000.00	204,000.00
QA/QC	34.00	75,000.00	100,000.00	87,500.00	2,550,000.00	3,400,000.00	2,975,000.00
Total		**227,000.00**	**326,000.00**	**276,500.00**	**7,718,000.00**	**11,084,000.00**	**9,401,000.00**

Figure 8.16 Economic analysis of landfill: closure costs (imperial)

Task	Area [sq m]	Costs [$/sq m]			Total Costs [$]		
		Min.	Max.	Average	Min.	Max.	Average
Final Grade Survey	13.76	7,413.16	14,826.32	11,119.74	102,000.00	204,000.00	153,000.00
Gas Management Layer	13.76	59,305.30	79,073.73	69,189.51	816,000.00	1,088,000.00	952,000.00
Compacted Clay Cap	13.76	64,247.41	126,023.76	95,135.58	884,000.00	1,734,000.00	1,309,000.00
Geomembrane Cap	13.76	44,478.97	56,834.24	50,656.61	612,000.00	782,000.00	697,000.00
Geocomposite	13.76	81,544.78	108,726.38	95,135.58	1,122,000.00	1,496,000.00	1,309,000.00
Cover and Vegetative Soil	13.76	32,123.70	64,247.41	48,185.55	442,000.00	884,000.00	663,000.00
Seed, Mulch, and Fertilizer	13.76	2,471.05	4,942.11	3,706.58	34,000.00	68,000.00	51,000.00
Gas Management System	13.76	71,660.57	86,486.89	79,073.73	986,000.00	1,190,000.00	1,088,000.00
Run-Off Control System	13.76	12,355.27	17,297.38	14,826.32	170,000.00	238,000.00	204,000.00
QA/QC	13.76	185,329.06	247,105.41	216,217.23	2,550,000.00	3,400,000.00	2,975,000.00
Total		**560,929.27**	**805,563.63**	**683,246.45**	**7,718,000.00**	**11,084,000.00**	**9,401,000.00**

Figure 8.17 Economic analysis of landfill: closure costs (metric)

Task	Area [acre]	Costs [$/acre] per Year			Total Costs [$] per Year		
		Min.	Max.	Average	Min.	Max.	Average
Security and Fencing Repair	34.00	3.00	6.00	4.50	102.00	204.00	153.00
Cap and Cover Maintenance	34.00	300.00	567.00	433.50	10,200.00	19,278.00	14,739.00
Leachate Machinery Maintenance	34.00	900.00	1,200.00	1,050.00	30,600.00	40,800.00	35,700.00
Landfill Gas Machinery Maintenance	34.00	450.00	570.00	510.00	15,300.00	19,380.00	17,340.00
Wells/Probes	34.00	20.00	30.00	25.00	680.00	1,020.00	850.00
Environmental Monitoring	34.00	450.00	575.00	512.50	15,300.00	19,550.00	17,425.00
Total (per Year)		2,123.00	2,948.00	2,535.50	72,182.00	100,232.00	86,207.00
Total (30 Years)		**63,690.00**	**88,440.00**	**76,065.00**	**2,165,460.00**	**3,006,960.00**	**2,586,210.00**

Figure 8.18 Economic analysis of landfill: postclosure and maintenance costs
(imperial)

Task	Area [sq m]	Costs [$/sq m] per Year			Total Costs [$] per Year		
		Min.	Max.	Average	Min.	Max.	Average
Security and Fencing Repair	13.76	7.41	14.83	11.12	102.00	204.00	153.00
Cap and Cover Maintenance	13.76	741.32	1,401.09	1,071.20	10,200.00	19,278.00	14,739.00
Leachate Machinery Maintenance	13.76	2,223.95	2,965.26	2,594.61	30,600.00	40,800.00	35,700.00
Landfill Gas Machinery Maintenance	13.76	1,111.97	1,408.50	1,260.24	15,300.00	19,380.00	17,340.00
Wells/Probes	13.76	49.42	74.13	61.78	680.00	1,020.00	850.00
Environmental Monitoring	13.76	1,111.97	1,420.86	1,266.42	15,300.00	19,550.00	17,425.00
Total (per Year)		5,246.05	7,284.67	6,265.36	72,182.00	100,232.00	86,207.00
Total (30 Years)		**157,381.43**	**218,540.02**	**187,960.73**	**2,165,460.00**	**3,006,960.00**	**2,586,210.00**

Figure 8.19 Economic analysis of landfill: postclosure and maintenance costs (metric)

8.3.2 Economic Analysis of WTE

General	
Price Level Ratio of Purchasing Power Parity Conversion Factor (Croatia)	0.6
Lifetime [years]	35
Yearly Waste Capacity [t]	100,000
Total Waste Capacity (35 Years) [t]	3,500,000
Steam Generator Efficiency [%]	80
Percentage of Plastic in MSW [%]	12.8
Percentage Heat of Produced Energy [%]	75
Percentage Electricity of Produced Energy [%]	25
Energy per Cubic Meter in Gas [kWh]	12.5
Average Lower Heating Value MSW [MJ/kg]	14.98
Average Lower Heating Value Plastic [MJ/kg]	36.16
Cost/Revenue per kWh Electricity [$/kWh]	0.1027
Cost/Revenue per Cubic Feet Gas [$/cubic ft]	0.0039
Cost/Revenue per Cubic Meter Gas [$/cubic meter]	0.1377

Figure 8.20 Economic analysis of waste-to-energy plant: assumptions

Type of Costs	Costs Croatia [€]	PP Conversion Factor	Euro-Dollar-Rate	Costs USA [$]
Infrastructure and Waste Storage	4,600,000.00			9,296,569.33
Combustion System and Steam Generator	19,500,000.00			39,409,370.00
Water and Steam System	8,000,000.00			16,167,946.67
Design	2,000,000.00			4,041,986.67
Construction	7,000,000.00	0.60	1.21	14,146,953.33
Electro-Mechanical Installations	5,000,000.00			10,104,966.67
Semi-Dry Treatment	1,200,000.00			2,425,192.00
Bag Filter	2,200,000.00			4,446,185.33
Selective Non-Catalytic Reduction (SNCR) System	800,000.00			1,616,794.67
Other Investment Costs	6,000,000.00			12,125,960.00
Total	4,200,000.00			113,781,924.67

Figure 8.21 Economic analysis of waste-to-energy plant: investment costs

Personnel Salaries	Number per Shift	Cost per Person per Year [$]	Shifts per Day	Total Costs [$]
Worker	15	30,000.00	3	1,350,000.00
Environmental Engineer	5	90,000.00	3	1,350,000.00
Maintenance Engineer	3	65,000.00	3	585,000.00
Manager (Energy Engineering)	1	95,000.00	3	285,000.00

Machine and Emission Costs	Costs Croatia [€]	PP Conversion Factor	Euro-Dollar-Rate	Costs USA [$]
System Maintenance (3% of Invest.)	1,689,000.00			3,413,457.74
Process Water	12,000.00			24,251.92
Natural Gas	85,000.00			171,784.43
Reagent for SNCR	80,000.00			161,679.47
Reagent for Semi-Dry Treatment	70,000.00	0.60	1.21	141,469.53
Bottom Ash-Disposal	1,380,000.00			2,788,970.80
Flying Ash from Steam Generator	138,000.00			278,897.08
Bag Filter Residues	1,575,000.00			3,183,064.50
Emission Fees	13,560.00			27,404.67

Type of Cost	Consumption per Ton of Waste [MWh/t]	Consumption per Year [MWh]	Costs per MWh [$/MWh]	Total Costs [$]
Electricity	0.10	10,000.00	102.70	102,700.00
Heat	0.05	5,000.00	11.02	2,754.55

Yearly Operating and Maintenance Costs [$]				13,866,434.69
Total Operating and Maintenance Costs (35 Years) [$]				485,325,214.20

Figure 8.22 Yearly operating and maintenance costs [$]

Type of Plastic	LHV [MJ/kg]	% in Waste [%]	Total [MJ/kg]
PET	23.2	14.16	3.29
HDPE	44.6	17.39	7.76
PP	42.7	22.62	9.66
LDPE	42.2	22.84	9.64
PVT	19.2	6.64	1.27
PS	42.0	2.71	1.14
Others	25.0	13.63	3.41
Total [MJ/kg]			36.16

Figure 8.23 Economic analysis of waste-to-energy plant: average lower heating value (LHV) of plastic

Type of Waste	LHV [MJ/kg]	% in Waste [%]	Total [MJ/kg]
Paper	19.12	25.00	4.78
Glass	0.00	4.20	0.00
Metals	0.00	9.40	0.00
Plastics	36.16	13.20	4.77
Rubber and Leather	31.28	3.40	1.06
Textiles	16.05	6.30	1.01
Wood	11.63	6.70	0.78
Food	6.05	15.20	0.92
Yard Trimmings	6.98	13.10	0.91
Other	21.05	3.50	0.74
Total [MJ/kg]			14.98

Figure 8.24 Economic analysis of waste-to-energy plant: average lower heating value (LHV) of municipal solid waste

State	Number of WTE Plants	Average WTE Tipping Fee [$/t]	Total
Alabama	1	25.00	25.00
Connecticut	7	64.00	448.00
Florida	12	52.92	635.04
Iowa	1	64.00	64.00
Massachusetts	7	69.00	483.00
Minnesota	9	55.00	495.00
New Hampshire	2	69.00	138.00
New Jersey	5	85.00	425.00
New York	10	72.34	723.40
Washington	3	98.00	294.00
Wisconsin	2	51.00	102.00
Total	59		3,832.44
Overall Average Tipping Fee			64.96

Figure 8.25 Economic analysis of waste-to-energy (WTE) plant: tipping fee

8.3.3 Economic Analysis of Recycling

Percentage of PET in Plastic Waste [%]	14.16
Average Price of Recycled PET Pellets [$/lb]	0.58
Price of Recycled PET Pellets [$/kg]	1.26
Electricity Price [$/kWh]	0.1027
Diesel Price [$/gallon]	2.198
Diesel Price [$/l]	0.5807
Water Price [$/gallon]	0.015
Water Price [$/l]	0.0040

Figure 8.26 Economic analysis of plastics recycling: overall assumptions

Lifetime [years]	10
Processing Building Size (sq. ft)	80,000
Waste Handled per Hour [t]	30
Yearly Waste Handling [t]	120,000
Total Waste Handling (10 years) [t]	1,200,000
Residues Rate [%]	2
Correctness of Manual Plastic Sorting [%]	91
Shifts per Day	2.0
Hours per Shift [h]	8.5
Breaktime per Shift [h]	0.5
Effective Working Time [h]	8.0
Work Days per Week [days]	5.0
Weeks per Year [weeks]	52.0
Yearly Working Hours [h]	4,160.0

Figure 8.27 Economic analysis of plastics recycling: materials recovery facility (MRF) assumptions

Building and Site	Initial Investment
Land [$]	675,000.00
Site Work [$]	720,000.00
Scale House [$]	600,000.00
MRF Building [$]	8,600,000.00
Construction, Planning & Surveying [$]	3,500,000.00
Machines	**Initial Investment**
Metering Bin [$]	150,000.00
OCC Screen [$]	175,000.00
Debris Roll Screen [$]	220,000.00
Newspaper Screen [$]	400,000.00
Polishing Screen [$]	280,000.00
Recycling Magnets [$]	35,000.00
Eddy Current Separator [$]	128,000.00
Optical Plastic Sorting Machine [$]	225,000.00
Baler [$]	550,000.00
Additional Equipment	**Initial Investment**
Conveyor [$]	50,000.00
Rolling Stock [$]	350,000.00
Collection Cars [$]	1,000,000.00
Total Investment Costs [$]	**17,658,000.00**

Figure 8.28
Economic analysis of plastics recycling: materials recovery facility (MRF) investment costs

Personnel Salaries per Year	Number per Shift	Cost per Person per Year [$]	Shifts per Day	Total Costs per Year [$]
Plant Manager	1	89,000.00		178,000.00
Operations Foreman	2	55,000.00		220,000.00
Sorters	25	30,000.00	2	1,500,000.00
Scale House Attendant	2	37,500.00		150,000.00
Equipment Operators	4	35,000.00		280,000.00
Spotters on Tip Floor	2	31,000.00		124,000.00
Marketing Manager	2	50,000.00	1	100,000.00

Facility Costs	Costs per Year [$]
Consumables/Services	28,000.00
Baling Wire	250,000.00
Insurance	45,000.00
Administration	20,000.00

Machines - Operating Costs	Efficiency [%]	Electricity Consumption [kW]	Running Hours per Year [h]	Electricity Costs per kWh [$/kWh]	Yearly Electricity Costs[$]
Metering Bin	100.00	15.00			6,215.04
OCC Screen	70.00	8.50			3,521.86
Debris Roll Screen	97.00	30.00			12,430.08
Newspaper Screen	85.00	5.50			2,278.85
Polishing Screen	91.00	10.00	4160	0.0996	4,143.36
Recycling Magnets	98.00	4.00			1,657.34
Eddy Current Separator	97.00	9.00			3,729.02
Optical PET Sorting Machine	98.00	13.00			5,386.37
Baler	100.00	63.00			26,103.17

Machines - Maintenance Costs	Maintenance Costs per Year [$]
Metering Bin	100.00
OCC Screen	10,000.00
Debris Roll Screen	10,000.00
Newspaper Screen	13,000.00
Polishing Screen	10,000.00
Recycling Magnets	5,000.00
Eddy Current Separator	5,000.00
Optical PET Sorting Machine	5,000.00
Baler	5,000.00

Conveyor - Operating Costs	Efficiency [%]	Electricity Consumption [kW]	Running Hours per Year [h]	Electricity Costs per kWh [$/kWh]	Yearly Electricity Costs[$]
Conveyor	100.00	5.60	4,160	0.1027	2392.50

Conveyor - Maintenance Costs	Maintenance Cost per Year [$]
Conveyor	10,000.00

Rolling Stock - Operating Costs	Efficiency [%]	Diesel Consumption [L/t]	Waste per Year [t]	Diesel Price [$/L]	Yearly Diesel Costs [$]
Rolling Stock	100.00	10.00	120,000	0.7434	892,056.61

Rolling Stock - Maintenance Costs	Maintenance Cost per Year [$]
Rolling Stock	5,000.00

Residues Costs	Percentage of Residues [%]	Residues per Year [t]	Tipping Fee per Ton [$]	Total Costs per Year [$]
Residues Costs per Year	2	2,400	40	96,000.00

Transportation and Collection Costs	Tons Collected per Year	Transportation Costs per Ton [$/t]	Transportation Costs per Year [$]
Transportation and Collection	120,000	36.84	4,420,800.00

Yearly Operating and Maintenance Costs [$]	8,254,465.90
Total Operating and Maintenance Costs (10 Years) [$]	82,544,659.03

Figure 8.29 Economic analysis of plastics recycling: material recovery facility (MRF) operating and maintenance costs

Lifetime [years]	10
Processing Building Size (sq. ft)	70,000
Yearly PET Capacity [t]	15,000
Total Waste Capacity (10 Years) [t]	150,000
Separation Efficiency [%]	91
Dollar - Pound Rate [$/£]	1.46
Shifts per Day	3.0
Hours per Shift [h]	8.5
Breaktime per Shift [h]	0.5
Effective Working Time [h]	8.0
Work Days per Week [Days]	5.0
Weeks per Year [Weeks]	52.0
Yearly Working Hours [h]	6,240.0

Figure 8.30 Economic analysis of plastics recycling: plastic recycling facility (PRF) assumptions

Building and Site	Initial Investment [£]	Dollar - Pound Rate [$/£]	Initial Investment
Design and Project Management	800,000.00	1.46	1,168,000.00
Civil Engineering	1,200,000.00		1,752,000.00
Land [$]			590,000.00
Site Work [$]			630,000.00

Machines	Initial Investment [£]	Dollar - Pound Rate [$/£]	Initial Investment [$]
Bale Breaker	5,620,000.00		8,205,200.00
Flake Washing Plant (Includes Granulating, Drying, Washing)	9,444,000.00	1.46	13,788,240.00
PET Extrusion	2,643,000.00		3,858,780.00
Metering Bin			150,000.00

Additional Equipment			Initial Investment [$]
Rolling Stock [$]			150,000.00

Total Investment Costs [$]			30,142,220.00

Figure 8.31 Economic analysis of plastics recycling: plastic recycling facility (PRF) investment costs

Personnel Salaries per Year	Number per Shift	Cost per Person per Year [$]	Shifts per Day	Total Costs [$]
Sorting and Bale Breaking	6	30,000.00		540,000.00
Flake Washing	2	30,000.00		180,000.00
PET Extrusion	2	30,000.00		180,000.00
Waste Processing	2	30,000.00	3	180,000.00
Equipment Operators	4	35,000.00		420,000.00
Maintenance Staff	2	62,715.00		376,290.00
Plant Manager	1	89,000.00		267,000.00
Marketing Manager	1	50,000.00	1	50,000.00

Machines - Maintenance Costs	Investment Costs [$]		Maintenance [%]	Maintenance Costs [$]
Maintenance Costs of Machines	25,852,220.00		5	1,292,611.00

Machines - Operating Costs	Electricity Consumption [kWh/t]	Electricity Price [$/kWh]	Water Consumption [L/t]	Water Price [$/l]	Tons per Year [t]	Total Costs [$]
Bale Breaker	8.00		0.00			12,324.00
Flake Washing Plant	350.00	0.1027	1.80	0.0040	15,000	539,281.99
PET Extrusion	325.00		0.40			500,686.28
Metering Bin	1.50		0.00			2,310.75

Rolling Stock - Operating Costs	Diesel Consumption [L/t]	Waste per Year [t]	Diesel Price [$/L]	Yearly Diesel Costs [$]
Rolling Stock	10.00	15,000	0.7434	111,507.07

Yearly Operating and Maintenance Costs [$]	4,652,011.09
Total Operating and Maintenance Costs (10 Years) [$]	46,520,110.91

Figure 8.32 Economic analysis of plastics recycling: plastic recycling facility (PRF) operating and maintenance costs

8.3.4 Optimization I: Reduction of Sorting Processes

Percentage of PET in Plastic Waste [%]	14.16
Average Price of Recycled PET Pellets [$/lb]	0.58
Price of Recycled PET Pellets [$/kg]	1.28
Electricity Price [$/kWh]	0.1027
Diesel Price [$/gallon]	2.814
Diesel Price [$/l]	0.7434
Water Price [$/gallon]	0.015
Water Price [$/l]	0.0040
Lifetime [years]	10
Processing Building Size (sq. ft)	80,000
Yearly Working Hours [h]	6,240
Yearly Plastic Waste Handling [t]	100,000
Total Plastic Waste Capacity (10 Years) [t]	1,000,000
Yearly PET Handling [t]	15,000
Total Waste Capacity (10 Years) [t]	150,000
Separation Efficiency [%]	91
Dollar - Pound Rate [$/£]	1.46
Shifts per Day	3.0
Hours per Shift [h]	8.5
Breaktime per Shift [h]	0.5
Effective Working Time [h]	8.0
Work Days per Week [days]	5.0
Weeks per Year [weeks]	52.0
Yearly Working Hours [h]	6,240.0

Figure 8.33 Analysis of optimization I: assumptions

Investment Costs Additional Building and Site	Initial Investment [$]
Additional Land [$]	85,000.00
Additional Site Work [$]	90,000.00
Scale House [$]	600,000.00
Additional Building [$]	1,000,000.00

Investment Costs Additional Machines	Initial Investment [$]
Metering Bin [$]	150,000.00
Optical PET Sorting Machine [$]	225,000.00
Baler [$]	550,000.00

Investment Costs Additional Equipment	Initial Investment [$]
Conveyor [$]	50,000.00
Additional Rolling Stock [$]	200,000.00
Collection Cars [$]	1,000,000.00
Total Additional Investment Costs [$]	**3,950,000.00**

Figure 8.34

Economic analysis of optimization I: additional investment costs

Personnel Salaries per Year	Number per Shift	Cost per Person per Year [$]	Shifts per Day	Total Costs [$]
Sorters	6	30,000.00		540,000.00
Scale House Attendant	2	37,500.00	3	225,000.00
Equipment Operators	1	35,000.00		105,000.00
Spotters on Tip Floor	1	31,000.00		93,000.00

Facility Costs per Year	Costs per Year [$]
Baling Wire	250,000.00

Machines - Operating Costs	Efficiency [%]	Electricity Consumption [kW]	Running Hours per Year [h]	Electricity Costs per kWh [$/kWh]	Yearly Electricity Costs[$]
Metering Bin	100.00	15.00			9,612.72
Optical PET Sorting Machine	98.00	13.00	6,240	0.1027	8,331.02
Baler	100.00	63.00			40,373.42

Machines - Maintenance Costs	Maintenance Costs per Year [$]
Metering Bin	100.00
Optical PET Sorting Machine	5,000.00
Baler	5,000.00

Rolling Stock - Operating Costs	Efficiency [%]	Diesel Consumption [L/t]	Waste per Year [t]	Diesel Price [$/L]	Yearly Diesel Costs [$]
Rolling Stock	100.00	10.00	100,000	0.7434	743,380.51

Rolling Stock - Maintenance Costs	Maintenance Costs per Year [$]
Rolling Stock	5,000.00

Transportation and Collection Costs	Tons per Year [t]	Transportation Costs per Ton [$/t]	Transportation Costs per Year
Transportation and Collection	100,000	36.84	**3,684,000.00**

Yearly Operating and Maintenance Costs [$]	**5,713,797.67**
Total Operating and Maintenance Costs (10 Years) [$]	**57,137,976.74**

Figure 8.35 Economic analysis of optimization I: additional operating and maintenance costs

8.3.5 Optimization II: Upcycling of Plastic Waste by Blending

Additional Investment Costs					Initial Investment [$]
LDPE-PP Optical Sorter					900,000.00

Additional Operating Costs	Efficiency [%]	Electricity Consumption [kW]	Running Hours per Year [h]	Electricity Costs per kWh [$/kWh]	Yearly Electricity Costs [$]
LDPE-PP Optical Sorter	95.00	40.00	4160	0.1027	17,089.28

Additional Maintenance Costs					Maintenance Costs per Year [$]
LDPE-PP Optical Sorter					10,000.00

Overall Additional MRF Costs [$]	1,170,892.80
Additional MRF Costs per Ton [$/t]	0.98

Figure 8.36 Economic analysis of optimization II for materials recovery facility (MRF): additional costs

Building and Site	Initial Investment [£]	Dollar - Pound Rate [$/£]	Initial Investment [$]
Design and Project Management	500,000.00	1.46	730,000.00
Land [$]	590,000.00		
Site Work [$]	630,000.00		

Machines	Initial Investment [£]	Dollar - Pound Rate [$/£]	Initial Investment [$]
Bale Breaker	5,620,000.00	1.46	8,205,200.00
Flake Washing Plant (Includes Granulating, Drying, Washing)	9,444,000.00		13,788,240.00
LDPE-PP Extruder	1,422,000.00		2,076,120.00
Metering Bin	150,000.00		

Additional Equipment			Initial Investment [$]
Rolling Stock			50,000.00

Total Investment Costs [$]	26,069,560.00

Figure 8.37 Economic analysis of optimization II for plastic recycling facility (PRF): additional investment costs

Personnel Salaries per Year	Number per Shift	Cost per Person per Year [$]	Shifts per Day	Total Costs [$]
Sorting and Bale Breaking	6	30,000.00		540,000.00
Flake Washing	2	30,000.00		180,000.00
LDPE-PP Extrusion	2	30,000.00	3	180,000.00
Waste Processing	2	30,000.00		180,000.00
Equipment Operators	4	35,000.00		420,000.00

Machines - Maintenance Costs	Investment costs [$]		Maintenance [%]		Maintenance Costs [$]
	24,069,560.00		5		1,203,478.00

Machines - Operating Costs	Electricity Consumption [kWh/t]	Electricity Price [$/kWh]	Water Consumption [L/t]	Water Price [$/l]	Tons per Year [t]	Total Costs [$]
Bale Breaker	8.00		0.00			12,324.00
Flake Washing Plant	350.00	0.1027	1.80	0.0040	15,000	539,281.99
LDPE-PP Extrusion	275.00		0.40			423,661.28
Metering Bin	1.50		0.00			2,310.75

Rolling Stock - Operating Costs	Diesel Consumption [L/t]	Waste per Year [t]		Diesel Price [$/L]		Yearly Diesel Costs [$]
Rolling Stock	10.00	15,000		0.7434		111,507.08

Yearly Operating and Maintenance Costs [$]		3,792,563.09
Total Operating and Maintenance Costs (10 Y.) [$]		37,925,630.91

Figure 8.38 Economic analysis of optimization II for plastic recycling facility (PRF): additional operating and maintenance costs

Index